Atmosphere, Weather, and Climate

Frontispiece. A photograph of almost the entire disc of the earth (more than ⅓ of the total earth's surface) taken from 35,800 km (about 22,000 miles) above the equator on November 10, 1967 by the U.S. Applications Technology Satellite (ATS-3). This shows depressions over the North (B) and South (K) Atlantic, from which cold fronts (A and L) extend equatorward. The Azores high-pressure area (C) is mostly cloud free, but convective cloud obscures some of the West Indian islands (E), and there are isolated thunderstorms in the South Atlantic (H). The cloud-covered equatorial trough (D) is clearly visible, the lower Amazon Basin is completely obscured by precipitation, and jet-stream clouds extend south-east from Brazil (G). The windward side of the central Andes is bathed in cloud (I), and there is much low cloud off the west coast of South America associated with the cold ocean current (NASA photograph).

Atmosphere, Weather, and Climate

R. G. BARRY and R. J. CHORLEY

HOLT, RINEHART AND WINSTON, INC.

New York Chicago San Francisco Atlanta Dallas

Preface

The rapid advances over the past ten to fifteen years in our understanding of atmospheric processes and global climates make a continual reappraisal of teaching methods and the content of textbooks essential. The traditional view of climatology as mere "bookkeeping" has at long last been abandoned by the majority of those interested in investigating the basic mechanisms of climatic differentiation, but the approaches of synoptic and dynamic climatology (dynamic climatology is concerned with the physical and dynamic explanation of atmospheric circulation patterns based on generalizations of meteorological data, whereas synoptic climatology provides interpretations of local or regional climates with reference to the large-scale circulation) have in general not yet found their way beyond the scientific papers into elementary textbooks.

The authors' aim is to help to fill this gap, particularly for those studying weather and climate in introductory courses of college or university Geography departments. At the same time, students in related disciplines such as agriculture, ecology and hydrology, and indeed all who are interested in the atmosphere and its weather, should find a basic nonmathematical introduction to modern ideas in this field in the present book. Some of the concepts which are introduced undoubtedly go rather beyond the general scope of courses at the levels mentioned, so that the book should also serve as a foundation for more advanced study. A guide to further reading is provided by the Bibliography. No attempt is made to present a comprehensive coverage of regional climates, but, by an examination of the weather and climate of the mid-latitudes of the northern hemisphere and the tropics in terms of a variety of themes, it is hoped to give the reader sufficient appreciation of climatic controls to apply these ideas elsewhere himself.

The first three chapters deal with the nature of the atmosphere—its energy budget, moisture balance and motion. The fourth chapter discusses air masses and the processes which lead to the development of frontal and other depressions. These basic concepts, together with such additional ones as are required, are then used to examine the climatic characteristics of mid-latitudes and the tropics. The book concludes with a brief consideration of the modifications of climate produced by the urban and forest environments and of the inherent variability of climate with time. A brief summary of the major schemes of climatic classification is given for reference purposes in Appendix I. It is worth emphasizing that the distinction between weather and climate is arbitrary. Average climatic conditions can be specified for particular places and time-periods, but every individual element of climate varies continuously in space and time. This fundamental point underlies the philosophy of the book: climate can only be understood through a knowledge of the workings of the atmosphere.

R. G. BARRY,
Institute of Arctic and Alpine Research,
University of Colorado,
Boulder, Colorado

R. J. CHORLEY,
Department of Geography,
University of Cambridge,
England

Acknowledgments

This book developed from an original manuscript by R. J. Chorley and A. J. Dunn, and the present authors wish to record their appreciation of Mr. Dunn's important contribution to the earlier draft.

The authors are also very much indebted to Dr. F. Kenneth Hare of Birkbeck College, London, now at the University of Toronto, Ontario, for his thorough and authoritative criticism of the preliminary text and his valuable suggestions for its improvement; also to Mr. Alan Johnson of Barton Peveril School, Eastleigh, Hampshire, for helpful comments on Chapters 1–3; and to Dr. C. Desmond Walshaw, formerly of the Cavendish Laboratory, Cambridge, and Mr. R. H. A. Stewart of the Nautical College, Pangbourne, for offering valuable criticisms and suggestions at an early stage in the preparation of the original manuscript. The authors accept complete responsibility for any remaining textual errors.

The figures were prepared by the cartographic and photographic staffs in the Geography Departments at Cambridge University (Mr. R. Coe, Miss R. King, Miss G. Seymour, and Mr. M. Young) and at Southampton University (Mr. A. C. Clarke, Miss B. Manning, and Mr. R. Smith).

Our grateful thanks go to our wives for their constant encouragement and forbearance.

The authors would like to thank the following learned societies, editors, publishers, organizations, and individuals for permission to reproduce figures, tables, and plates.

Learned Societies

American Geographical Society for Fig. 1.30 from the *Geographical Review*.
American Meteorological Society for Fig. 4.18 from the *Bulletin;* for Fig. 3.23 from the *Journal of Applied Meteorology;* and for Fig. 3.6 from the *Compendium of Meteorology*.
Geographical Association for Fig. 2.2 and Plates 23 and 24 from *Geography*.

Royal Meteorological Society for Figs. 1.1, 2.10, 2.13, 5.4, 5.5, 6.18, and 8.3 from the *Quarterly Journal;* for Fig. 8.4 from *World Climate 8000–0* B.C.; and for Figs. 1.13, 2.5, 3.7, 3.26, and 4.9 and Plates 15 and 17 from *Weather*.

Editors

Endeavour for Fig. 2.18.
Erdkunde for Figs. 1.23B, Appendix 1.1B and Appendix 1.2.
Geographical Reports of Tokyo Metropolitan University for Fig. 6.20.
Meteorologische Rundschau for Fig. 5.19.
Tellus for Figs. 5.6, 5.7, 6.10, and 6.16.

Publishers

Allen and Unwin, London, for Figs. 1.15 and 1.17B from *Oceanography for Meteorologists* by H. V. Sverdrup.

Cambridge University Press for Fig. 7.13 from *The Tropical Rain Forest* by P. W. Richards.

Cleaver-Hume Press, London, for Fig. 3.10 from *Realms of Water by* Ph. H. Kuenen.

The Controller, Her Majesty's Stationery Office (Crown Copyright Reserved) for Fig. 2.8 from *Geophysical Memoir No. 102-Average Water Vapour Content of the Air*, by J. K. Bannon and L. P. Steele; for Fig. 1.16 from *Meteorological Office Scientific Paper No. 6, M.O. 685* by F. E. Lumb; for Fig. 2.6 from *Ministry of Agriculture Technical Bulletin No. 4* by R. T. Pearl *et al.;* for Figs. 4.7, 4.8, and 4.10 from *A Course in Elementary Meteorology* by D. E. Pedgley; for Fig. 4.11 from *British Weather in Maps* by J. A. Taylor and R. A. Yates (published by Macmillan, London); and for the tephigram base of Fig. 2.14 from *RAF Form 2810*.

J. M. Dent and Sons Ltd, for Fig. 5.16 from *Canadian Regions* by D. F. Putnam (Ed.).

Folia Geographica Danica for Fig. 8.5 by L. Lysgaard.

Harvard University Press for Figs. 1.17A, 1.21, 7.6, 7.7, 7.9B, and 7.10A from *The Climate Near the Ground* (2d Ed.) by R. Geiger; Hutchinson, London, for Figs. 7.2, 7.3, and 7.5 from the *Climate of London* by T. J. Chandler.

Justus Perthes, Gotha, for Fig. 2.24 from *Petermann's Geographische Mitteilungen*, Jahrgang 95.

Macmillan and Co., London, for Fig. 4.11 from *British Weather in Maps* by J. A. Taylor and R. A. Yates.

McGraw-Hill Book Co., New York, for Fig. 2.21 from *Handbook of Meteorology* by F. A. Berry, E. Bollay, and N. R. Beers (Eds.); for Fig. 1.6 from *General Meteorology* by H. R. Byers; for Fig. 3.28 from *Dynamical and Physical Meteorology* by G. J. Haltiner and F. L. Martin; for Figs. 7.9A and 7.10B from *Forest Influences* by J. Kittredge; for Fig. 2.22 from *Water-Resources Engineering* by R. K. Linsley and J. B. Franzini; for Figs. 2.9 and 2.15 from *Introduction to Meteorology* (1st Ed.) by S. Petterssen; for Figs. 2.17 and 3.15

from *Introduction to Meteorology* (2d Ed.) by S. Petterssen; for Fig. 6.3 from *Tropical Meteorology* by H. Riehl; for Figs. 5.12, 5.15, and 6.13 from *The Earth's Problem Climates* by G. T. Trewartha; and for Fig. 1.29 from *Handbook of Geophysics and Space Environments*, edited by Shea L. Valley (published by McGraw-Hill Book Company).

Charles E. Merrill Publishing Company for Fig. 1.23A from *Meteorology* by A. Miller.

Methuen and Co. Ltd, London for Figs. 2.1, 3.25, and 3.27 from *Models in Geography* by R. J. Chorley and P. Haggett (eds.).

North-Holland Publishing Company, Amsterdam, for Fig. 2.23 from "A Moisture Balance Profile of the Sierra Nevada" by C. F. Armstrong and C. K. Stidd in *Journal of Hydrology*, vol. 5, page 268.

Oxford University Press for Fig. 1.8 from *Exploring the Atmosphere* by G. M. B. Dobson.

Pergamon Press, London and Oxford, for Fig. 7.4 from *Atmospheric Pollution* by A. R. Meetham; and for Plate 9 from *Hydrometeorology* by J. P. Bruce and R. H. Clark.

Pitman, London, for Fig. 3.19 from *Tropical and Equatorial Meteorology* by M. A. Garbel.

Princeton University Press for Figs. 5.16 and 5.17 from *The Moisture Balance* by C. W. Thornthwaite and J. R. Mather (Publications in Climatology, Centerton, New Jersey) and for Fig. 7.1 from *Design with Climate* by V. Olgyay.

Scientific American Inc., New York, for Figs. 1.2 and 1.3 by G. N. Plass; and for Fig. 1.24 by R. E. Newell.

Springer-Verlag, Vienna/New York, for Fig. 1.31 from *Meteorologische Rundschau*, 2(1968), page 48; and for Fig. 2.23 from *Archiv für Meteorologie, Geophysik und Bioklimatologie*, B, 15(1967), page 235 by S. L. Hastenrath.

University of California Press for Fig. 6.5 and Plate 27 from *Cloud Structure and Distributions over the Tropical Pacific Ocean* (1964) by J. S. Malkus and H. Riehl.

The University of Chicago Press for Figs. 1.9, 1.25, 2.4 and for Table 7.1 from *Physical Climatology* by W. D. Sellers.

The University of Wisconsin Press for Fig. 6.20 from *The Earth's Problem Climates* (1961) by G. T. Trewartha.

Van Nostrand Reinhold Company for Fig. 6.22 from *Encyclopedia of Atmospheric Sciences and Astrogeology* (1967) by R. W. Fairbridge.

Walter De Gruyter and Co., Berlin, for Fig. 5.1 from *Allgemeine Klimageographie* by J. Blüthgen.

John Wiley and Sons, Inc., New York, for Fig. 4.5 from *Physical Geography* (1st Ed.) by A. N. Strahler; for Fig. 1.18 from *Physical Geography* (2d Ed.) by A. N. Strahler; for Fig. 8.2 from *Physical Geography* (3d Ed.) by A. N. Strahler; for Figs. 1.7, 2.12, and 3.13 from *Introduction to Physical Geography* by A. N. Strahler; and for Fig. 1.10 from *Meteorology, Theoretical and Applied* by E. W. Hewson and R. W. Longley.

Organizations

Deutscher Wetterdienst, Zentralamt, Offenbach am Main, for Fig. 6.19.

Environmental Science Services Administration ESSA for Plates 4, 5, 8, 11, 12, 14, 16, 18, 23, and 30.

National Aeronautics and Space Administration NASA for the Frontispiece and for Plates 1, 2, 10, 19, 22, 25, 26, 28, and 29.

New Zealand Meteorological Service, Wellington, New Zealand, for Figs. 6.15, 6.17, and 6.21 from the *Proceedings of the Symposium on Tropical Meteorology* by J. W. Hutchings (Ed.).

Press Association-Reuters Ltd., London, for Plate 3.

Quartermaster Research and Engineering Command, Natick, Massachusetts, U.S.A., for Fig. 5.11 by J. N. Rayner.

United Nations Food and Agriculture Organization, Rome, for Fig. 7.12B from *Forest Influences.*

United States Department of Agriculture, Washington D.C., for Figs. 7.11B and 7.12A from *Climate and Man.*

United States Weather Bureau for Figs. 2.20, 3.11, 3.16, 4.17, and Plate 13 from the *Monthly Weather Review;* and for Fig. 4.14 from *Research Paper No. 40.*

Individuals

Mr. C. F. Armstrong and Mr. C. K. Stidd of the Desert Research Institute, University of Nevada, for Fig. 2.23.

Dr. F. C. Bates, the National Science Foundation and the University of Kansas, for Plate 20.

Mrs. Ruth E. Chambers of the Department of Geography, University of Calgary, for Plate 6.

Dr. G. C. Evans of the Department of Botany, University of Cambridge, for Fig. 7.13A.

Dr. H. Flohn of the Meteorological Institute, University of Bonn, for Figs. 3.22 and 6.9.

Dr. S. Gregory of the Department of Geography, University of Sheffield, for Fig. 6.8.

Dr. S. L. Hastenrath of the Meteorology Department, University of Wisconsin, for Figs. 1.31 and 2.23.

Dr. L. H. Horn and Dr. R. A. Bryson of the Meteorology Department, University of Wisconsin, for Fig. 5.9.

Dr. F. H. Ludlam of Imperial College, London, for Plates 15 and 17.

Mr. Kiuo Maejima of Tokyo Metropolitan University for Fig. 6.20.

Mr. D. A. Richter, Analysis and Forecast Division, National Meteorological Center, for Fig. 4.17.

Dr. R. S. Scorer of Imperial College, London, and Mrs. Robert F. Symons for Plate 7.

Dr. P. A. Sheppard of Imperial College, London, for Plate 16.

Contents

1

Atmospheric Energy

The key to atmospheric processes is the radiant energy which the earth
and its atmosphere receive from the sun. In order to study the receipt of this
energy we need to begin by considering the nature of the atmosphere—its
composition and basic properties.

A. COMPOSITION OF THE ATMOSPHERE

1. Total atmosphere. Air is a mechanical mixture of gases, not a chemical
compound. It is highly compressible, such that its lower layers are very
much more dense than those above; for instance, the average density
decreases from about 1.2 kg/m³ at the surface to 0.7 kg/m³ at 5000 meters
(approximately 16,000 feet), the approximate limit of human habitation.
However, the considerable thickness of the atmosphere causes it to exert a
surface pressure sufficient to support a mercury column 760 mm (29.9
inches) high. This is equivalent to a force of just over 1 kg/cm² (14.7 lb/in²).

Although the atmosphere is composed of a number of gases, five of them—
nitrogen, oxygen, argon, carbon dioxide, and water vapor—make up 99.997%
of it by volume below 90 km. Rocket observations show that the atmospheric
gases are mixed in remarkably constant proportions up to at least 50 km
(approximately 30 miles) and, for example, surface samples of atmospheric
oxygen show insignificant variations over the globe. Table 1.1 gives some
idea of the average composition of dry air, in addition to which there are
significant quantities of water vapor, and *aerosols*—particles of smoke, dust,
and sea salt larger than molecular size.

Having made the above generalizations about the atmosphere, it is
immediately necessary to qualify them by drawing attention to the variations
which occur in the composition of the atmosphere when height, latitude, and
time are varied.

1

Table 1.1:   Average composition of the dry atmosphere below 25 km.

Component	Symbol	Volume % (dry air)	Molecular weight
Nitrogen	N₂	78.08	28.02
Oxygen	O₂	20.94	32.00
*‡Argon	Ar	0.93	39.88
Carbon dioxide	CO₂	0.03 (very variable)	44.00
‡Neon	Ne	0.0018	20.18
*‡Helium	He	0.0005	4.00
†Ozone	O₃	0.00006	48.00
Hydrogen	H	0.00005	2.02
‡Krypton	Kr	Trace	
‡Xenon	Xe	Trace	
Methane	Me	Trace	

* Decay products of potassium and uranium.
† Recombination of oxygen.
‡ Inert gases.

2. Variations with height. The lighter gases (hydrogen and helium especially) might be expected to become more abundant in the upper atmosphere, but large-scale turbulent mixing of the atmosphere prevents such diffusive separation even at heights of many tens of kilometers above the surface. The height variations which do occur are related to the source-locations of the two major nonpermanent gases—water vapor and ozone. Since both absorb some solar and terrestrial radiation the heat budget and vertical temperature structure of the atmosphere are considerably affected by the distribution of these two gases (see Chapter 1, Sections D.2 and G).

Water vapor comprises up to 4% of the atmosphere by volume (about 3% by weight) near the surface, but is almost absent above 10 to 12 km. It is supplied to the atmosphere by evaporation from surface water or by transpiration from plants and is transferred upwards by atmospheric turbulence. Turbulence is most effective below about 10 km (see Chapter 1, Section G.1) and as the maximum possible water vapor density of cold air is anyway very low (see Chapter 1, Section B.2), there is little water vapor in the upper layers of the atmosphere.

Ozone (O_3) is concentrated mainly between 15 to 35 km. The upper layers of the atmosphere are irradiated by ultraviolet radiation from the sun which causes the break-up of oxygen molecules in the layer between about 80 to 100 km (that is, $O_2 \rightarrow O + O$). These separated atoms ($O + O$) may then individually combine with other oxygen molecules to create ozone.

$$O_2 + O + M \rightarrow O_3 + M,$$

where M represents the energy and momentum balance provided by collision with a third atom or molecule. Such three-body collisions are rare at 80 to

100 km because of the very low density of the atmosphere, while below about 35 km most of the incoming ultraviolet radiation has already been absorbed at higher levels. Therefore ozone is mainly formed between 30 and 60 km where collisions between O and O_2 are more likely. Ozone itself is unstable and it may be destroyed either by collisions with monatomic oxygen to recreate oxygen (that is, $O_3 + O \rightarrow O_2 + O_2$) or by the action of radiation on it.

The constant metamorphosis of oxygen to ozone and from ozone back to oxygen by photochemical processes maintains an approximate equilibrium above about 40 km, but the ozone mixing ratio is a maximum at about 35 km, whereas maximum ozone density occurs lower down between 20 and 25 km.[1] This is the result of some circulation mechanism transporting ozone downwards to levels where its destruction is less likely, allowing an accumulation of the gas to occur. Even so it is essential to realize that, despite the importance of the ozone layer, if the atmosphere were compressed to sea level (at normal sea-level temperature and pressure) ozone would contribute only about 3 mm to the total atmospheric thickness of 8 km (Fig. 1.1).

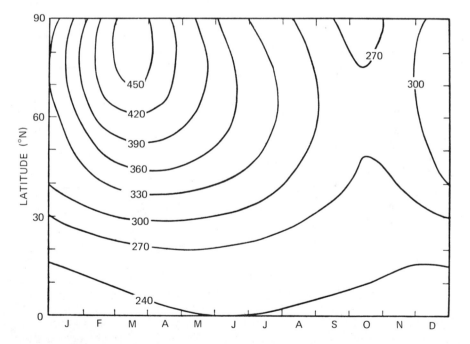

Fig. 1.1. The monthly variation of total atmospheric ozone by latitude in the northern hemisphere. The units are 10^{-3} cm of ozone at standard atmospheric temperature and pressure (from Godson, 1960).

---

[1] Mixing ratio = mass of ozone per unit mass of dry air. Density = mass per unit volume.

3. Variations with latitude and season. Variations of atmospheric composition with latitude and season are particularly important in the case of water vapor and ozone.

Ozone content is low over the equator and high over latitudes north of 50°N, especially in spring (Fig. 1.1). If the distribution were solely the result of photochemical processes the maximum would occur in June near the equator and the anomalous pattern must be due to a poleward transport of ozone. The movement is apparently from higher levels (30–40 km) in low latitudes towards lower levels (20–25 km) in high latitudes during winter months. Here the ozone is stored during the *polar night* giving rise to an ozone-rich layer in early spring. The type of circulation responsible for this transfer is not yet known with certainty, although it does not seem to be a simple, direct one. In the southern hemisphere there is a similar distribution pattern to that of Fig. 1.1 except that poleward of 55°S the maximum is later and less pronounced than in the northern hemisphere.

The water-vapor content of the atmosphere is closely related to air temperature (see Chapter 1, Section B.2 and Chapter 2, Sections A and B) and is therefore greatest in summer and in low latitudes. There are, however, obvious exceptions to this generalization, such as the tropical desert areas of the world.

The carbon dioxide content of the air (averaging about 315 parts per million) has a large seasonal range in higher latitudes in the northern hemisphere. At 50°N the concentration ranges from 310 ppm in late summer to 318 ppm in spring. The low summer values are related to the assimilation of CO_2 by the cold polar seas. Over the year a small net transfer of CO_2 from low latitudes to high latitudes takes place to maintain an equilibrium content in the air.

4. Variations with time. The quantities of carbon dioxide and ozone in the atmosphere may be subject to variations over a long time-period and these are of special significance because of their possible effect on the radiation budget.

Carbon dioxide (CO_2) enters the atmosphere mainly by the action of living organisms on land and in the ocean. The decay of organic elements in the soil and the burning of fossil fuels are additional minor sources (Fig. 1.2). It is obvious that if this production were not countered in some way the total quantity of carbon would steadily increase. A balance, or dynamic equilibrium, is maintained primarily by photosynthesis which removes approximately 3% of the world's total carbon dioxide annually. In the oceans the carbon dioxide ultimately goes to produce carbonate of lime, partly in the form of shells and the skeletons of marine creatures. On land the dead matter becomes humus which may subsequently form a fossil fuel. As Fig. 1.2 shows, the exchanges between the atmosphere and the other reservoirs are more or less balanced.

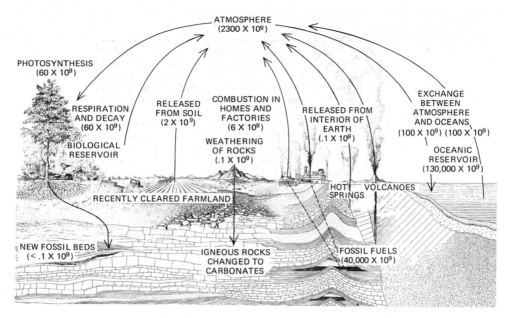

Fig. 1.2. The balance of carbon dioxide in the atmosphere. The numbers in parenthesis after each process involved in this exchange indicate the number of tons of carbon dioxide being absorbed or released each year (from Plass, 1959).

Yet this balance is not an absolute one, for between 1900 and 1935 the total quantity of atmospheric carbon dioxide is estimated to have risen by 9%, due, it is believed, to increased burning of fossil fuels. The calculated increase that combustion should cause, however, is almost twice this figure. These global estimates are based on observations made mainly in industrialized countries but, in spite of the resultant uncertainty attaching to the data, the discrepancy is sufficiently large to indicate that one of the major reservoirs of carbon dioxide, probably the ocean, is absorbing about half of the input to the atmosphere and thereby acting as a buffer to change. Carbon dioxide absorbs a great deal of terrestrial radiation which would be otherwise lost to space (see Fig. 1.8). An excess of carbon dioxide will therefore allow the atmosphere to tap the energy initially provided by the sun in larger quantities. The suggested link between carbon dioxide content and air temperatures is illustrated in Fig. 1.3, but it should be pointed out that the temperature rise has been more marked in polar latitudes than in the industrial countries of the world and that the temperature rise seems, according to some experts, to have terminated about 1940.

Ozone variations may occur if the solar output of ultraviolet radiation changes. Since ozone absorbs solar and terrestrial radiation the effects of any such variations may be complex. This idea forms the basis of a recent hypothesis of climatic change (see Chapter 8, Section C), although many

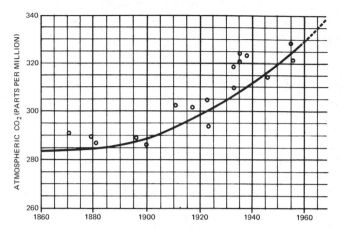

Fig. 1.3. Measurements showing (above) the rise of carbon dioxide content in the atmosphere during the past century, and (below) the general rise of mean annual temperatures (30-year moving means) at six stations during the past century (data compiled by G. S. Callendar. From Plass, 1959).

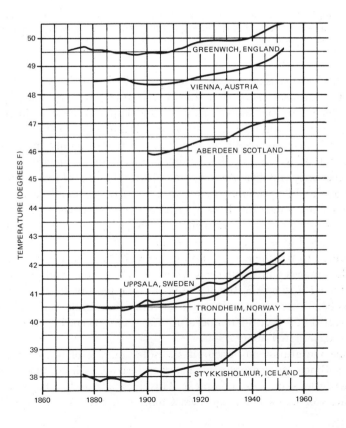

of the steps in the argument require detailed investigation before a definite theory can be put forward.

B. MASS OF THE ATMOSPHERE

It is now necessary to examine some of the mechanical laws which the atmospheric gases obey. Two simple laws specify the main factors governing changes in pressure. The first, Boyle's Law, states that, at a constant temperature, the volume (V) of a mass of gas varies inversely as its pressure (P), that is,

$$P = \frac{K_1}{V}$$

$(K_1$ is a constant); and the second, Charles's Law, that, at a constant pressure, volume varies directly with absolute temperature (T), that is,

$$V = K_2 T.$$

These laws imply that the three qualities of pressure, temperature, and volume are completely interdependent, such that any change in one of them will cause a compensating change to occur in one, or both, of the remainder. In the case of the atmosphere, however, it is convenient to use density $(\rho = \text{mass/volume})$ rather than volume. The gas laws may be combined to give the following relationship, known as the equation of state:

$$P = R\rho T,$$

where R is a gas constant (in this case for dry air).

1. Total pressure. As a result of the compressibility of the atmosphere, air density decreases strikingly with height, so that 50% of the total mass of air is found below 5 km (Fig. 1.4). Atmospheric pressure, depending as it does on the weight of the overlying atmosphere, decreases logarithmically with height.

Pressure is measured as a force per unit area. The units used by meteorologists are called millibars (mb), one millibar being equal to a force of 100 newtons acting on one square meter.[2] Pressure readings are made with a mercury barometer which in effect measures the weight of the column of mercury that the atmosphere is able to support in a vertical glass tube. The closed, upper end of the tube has a vacuum space and its open, lower end is immersed in a cistern of mercury. By exerting pressure downwards on the surface of mercury in the cistern, the atmosphere is able to support a mercury

---

[2] See Appendix 3.

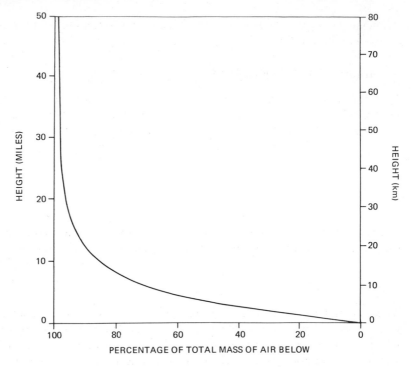

Fig. 1.4. The percentage of the total mass of the atmosphere lying below elevations up to 80 km (50 miles). This illustrates the shallow character of the earth's atmosphere.

column in the tube of about 762 mm (30 in or approximately 1016 mb). However, in order to compare the pressure at different geographical locations a further factor must be taken into account. A correction has to be made to mercury barometer readings for variations in pressure attributable to differences in gravity, which at sea level varies from 9.78 m/sec² at the equator to 9.83 m/sec² at the poles. Pressure readings are referred to the standard value of 9.81 m/sec² for 45° latitude. Mercury barometer readings must also be standardized to allow for the thermal expansion of mercury. The adopted standard temperature is 0°C (32°F).

Near the surface the rate of decrease of pressure with height is about one millibar per 10 meters. With increasing height, however, the drop in air density causes a decrease in this rate. The temperature of the air can affect this rate of pressure decrease, which is greater for cold dense air (see Chapter 3, Section C.4), although the relationship between pressure and height is so significant that meteorologists often express elevations in millibars, such that 1000 mb represents sea level, 500 mb about 5500 meters, and 300 mb about 9000 meters. A conversion nomogram for an idealized standard atmosphere is given in Appendix 2. Mean sea-level pressure is 1013.25 mb (equivalent to 14.7 lb/in²). On average, nitrogen contributes about 760 mb, oxygen 240 mb, and water vapor 10 mb. In other words, each gas exerts a partial pressure independent of the others.

2. Vapor pressure. At any given temperature there is a limit to the density of water vapor in the air, with a consequent upper limit to the vapor pressure. This is termed the *saturation vapor pressure* (e_s) and Fig. 1.5 illustrates how it increases with temperature, reaching a maximum of 1013 mb (one atmosphere) at boiling point. Attempts to introduce more vapor into the air when the vapor pressure is at saturation produce condensation of an equivalent amount of vapor. Fig. 1.5 shows that whereas the saturation vapor pressure has a single value at any temperature above freezing point, below 0°C the saturation vapor pressure above an ice surface is lower than that above a supercooled water surface. The significance of this will be discussed in Chapter 2, Section D.1.

Vapor pressure (e) varies with latitude and season from about 0.2 mb over northern Siberia in January to over 30 mb in the tropics in July, but this is not reflected in the pattern of surface pressure. Pressure decreases at the surface when some of the overlying air is displaced horizontally, and in fact the air in high-pressure areas is generally dry owing to dynamic factors,

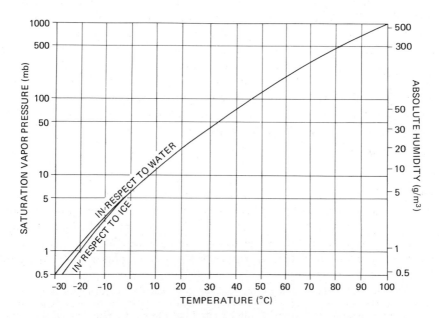

Fig. 1.5. Plot (semilogarithmic) of the saturation vapor pressure as a function of temperature (i.e. the dew-point curve). Below 0°C the atmospheric saturation vapor pressure is less with respect to an ice surface than with respect to a water drop. Thus, condensation may take place on an ice crystal at lower air humidity than is necessary for the growth of water drops.

particularly vertical air motion (see Chapter 3, Section C.5), while air in low-pressure areas is usually moist.

C. INSOLATION (ASSUMING NO ATMOSPHERE)

The prime source of the energy injected into our atmosphere is the sun, which is continually shedding part of its mass by radiating electromagnetic energy waves and high-speed particles into space. This constant emission, termed insolation, is important because it represents in the long run almost all the energy available to the earth (except for a small amount emanating from the radioactive decay of earth minerals). The amount of insolation received by the earth, assuming for the moment that there is no interference from the atmosphere, is affected by four factors.

1. Solar output. Of the total solar energy sent out into space the earth intercepts only some two thousand millionth, equivalent to a power of 1.8 $\times$ 10^{14} kW. The energy received on a surface normal to the solar beam is about 2 cal/cm²/min (1.396 kW/m²)[3]; this is termed the *solar constant*. The small proportion of solar energy available to the earth is reflected in the difference between the surface temperatures of the sun and earth; the temperature at the surface of the former is believed to be about 6000°K (°C = °K − 273°), whereas the mean temperature of the earth's atmosphere is only about 250°K (that is − 23°C) and that of the earth's surface is only 283°K (10°C or 50°F).[4] Fig. 1.6 shows the range of wavelengths (mostly short) of the emitted sun's energy, together with the longer-wave reradiation by the earth and its atmosphere. These curves are constructed on the assumption that the sun and the earth behave as *black bodies*. Such bodies absorb all energy falling on them and radiate energy at a rate directly proportional to the fourth power of their absolute temperature (that is, measured in °K). The radiant energy of a black body is emitted at a rate, F,

$$F = \sigma T^4, \qquad .817 \times 10^{-10} \; cal/cm^2/min$$

where σ = the Stefan-Boltzmann constant; this is known as Stefan's Law. At a given temperature the emission in each wavelength from a black body is the maximum possible. Most solids and liquids behave like black bodies, whereas gases do not. The wavelength of maximum emission (λ_{max}) varies inversely with the absolute temperature of the radiating body in the equation

---

[3] "Calorie" refers throughout to the small or gram-calorie. Another unit in common use is the Langley (ly) (1 ly/min = 1 cal/cm²/min). A calorie is the heat required to raise the temperature of 1 gram of water from 14.5° to 15.5°C. The units of the international metric system (kW/m²) are given in Appendix 3. At the present time the data in most references are still in calories.

[4] °C = degrees Celsius; °K = degrees Kelvin (or Absolute). Conversions for °C and °F are given in Appendix 2.

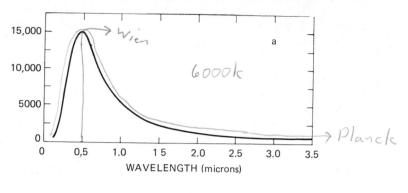

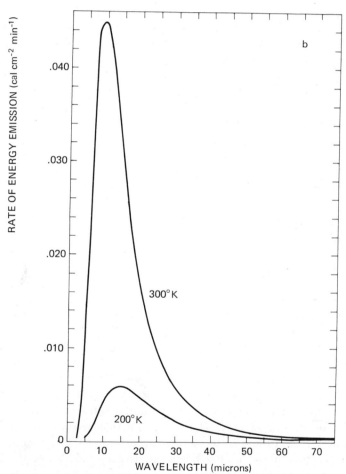

Fig. 1.6. Black-body emission curves for (a) the sun (6000°K) and (b) for bodies with temperatures of 300°K and 200°K. Note the contrasting vertical scales of emitted energy in units of cal/cm²/min and the contrasting wavelength scales between (a) and (b). The mean temperature of the earth's surface is 283°K. (from Byers, 1959).

expressing Wien's Law.

$$\lambda_{max} = \frac{2897}{T} \times 10^{-6} m.$$

Thus solar radiation is very intense and is mainly short wave between about 0.2 μ and 4.0 μ, with a peak in the middle part of the spectrum, whereas the much weaker terrestrial radiation has a peak intensity at about 10 μ and a range of about 4 μ to 100 μ (1 μ = 1 micron = 10^{-6} m). The solar constant has been variously estimated to be between 1.94–2.02 cal/cm²/min. The most recent measurements indicate a value of 1.36 kW/m² (1.95 cal/cm²/min).

The solar constant of sun's energy received by the earth's outer atmosphere was between 1905 and 1926 estimated as an average rate of 1.94 cal/cm²/min (or 1.94 ly/min), although more recent measurements indicated a value of 2.0 cal/cm²/min. The figure appears to undergo small periodic variations of 1 to 2% related to the sunspot cycle, but measurements of the solar constant are subject to errors of similar magnitude to the estimated fluctuations and their reality, therefore, is still doubtful. Nevertheless, undoubted variations do occur within the ultraviolet band of the spectrum, up to twenty times more ultraviolet radiation may be emitted at certain wavelengths during a sunspot maximum than during a sunspot minimum. However, no clear link between the 11-year sunspot cycle and weather variations has yet been demonstrated, in spite of many attempts to discover such a relationship. In the long term, assuming that the earth behaves as a black body, a long-continued difference of 2% in the solar constant could change the effective mean temperature of the earth's surface by as much as 1.2°C (2.2°F), and a 10% change might alter this temperature by as much as 6°C (10.7°F). The drop in surface temperature often experienced on a sunny day when a cloud temporarily cuts off the direct solar radiation illustrates our reliance upon the sun's radiant energy.

2. Distance from the sun. The ever-changing distance of the earth from the sun produces more frequent variations in our receipt of solar energy. Owing to the eccentricity of the earth's orbit round the sun, the receipt of solar energy on a surface normal to the beam is 7% more on January 3 at the perihelion than on July 4 at the aphelion. In theory (that is, discounting the interposition of the atmosphere and the difference in degree of conductivity between large land and sea masses) this difference should produce an increase in the effective January world surface temperatures of about 4°C (7°F), over those of July. It should also make northern hemisphere winters warmer than those in the southern, and southern hemisphere summers warmer than those in the northern. In practice, atmospheric heat circulation and the effects of continentality substantially mask this global tendency and the actual seasonal contrast between the hemispheres is reversed. Fig. 1.7 graphically illustrates the seasonal variations of energy receipt, with latitude. Actual amounts of insolation received on a horizontal

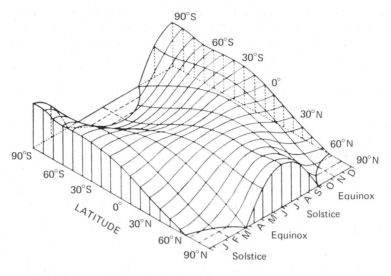

Fig. 1.7. The variations of insolation with latitude and season for the whole globe, assuming no atmosphere. This assumption explains the abnormally high amounts of insolation received at the poles in summer, when daylight lasts for 24 hours each day (after W. M. Davis. From Strahler, 1965).

surface outside the atmosphere are given in Table 1.2. The intensity on a horizontal surface (I_h) is determined from

$$I_h = I_0 \sin \mu,$$

where $I_0 = $ the solar constant and $\mu = $ the angle between the surface and the solar beam.

Table 1.2: Insolation on a horizontal surface outside the atmosphere, units: cal/cm²/day *(after K. Ya. Kondratiev)*.

Date	90°N	70	50	30	0	30	50	70	90°S
Dec 22	0	0	181	480	962	1073	1089	1114	1185
Feb 4	0	25	298	586	905	1003	937	809	834
Mar 21	0	316	593	799	923	799	593	316	0
May 6	796	722	894	958	863	560	285	24	0
June 22	1110	1043	1020	1005	814	450	170	0	0

3. Altitude of the sun. The altitude of the sun (that is, the angle between its rays and a tangent to the earth's surface at the point of observation) also affects the amount of insolation received at the surface of the earth. The greater the sun's altitude the more concentrated is the radiation intensity per unit area at the earth's surface. There are, in addition, important variations with solar altitude of the proportion of radiation reflected by the surface, particularly in the case of a water surface (see Chapter 1, Section D.5). The principal factors which determine the sun's altitude are, of course, the latitude of the site, the time of day, and the season.

4. Length of day. The length of daylight also affects the amount of insolation which is received. Obviously the longer the time during which the sun shines the greater is the quantity of radiation which a given portion of the earth will be able to receive.

The combination of all these factors produces the pattern of receipt of solar energy at the top of the atmosphere shown by Fig. 1.7. The polar regions receive their maximum amounts of insolation during their summer solstices, which is the period of continuous day. The amount of insolation received during the December solstice in the southern hemisphere is greater than that received by the northern hemisphere during the June solstice, due to the previously mentioned elliptical path of the earth around the sun. The equator has two insolation maxima at the equinoxes and two minima at the solstices, due to the apparent passage of the sun during its double annual movement between the northern and southern hemispheres.

D. SURFACE RECEIPT OF INSOLATION AND ITS EFFECTS

1. Energy transfer within the earth-atmosphere system. So far we have described the distribution of insolation as if all of it were available at the earth's surface. This is of course an unreal view because of the effect of the atmosphere on energy transfer. Heat energy can be transferred by the three following mechanisms:

(1) Radiation. Electromagnetic waves can transfer energy (both heat and light) between two bodies without the necessary aid of an intervening material medium. This is so with solar energy through space, whereas the earth's atmosphere only allows the passage of radiation at certain wavelengths and restricts that at others.

(2) Conduction. Under this mechanism the heat passes through a substance from point to point by means of the transfer of adjacent molecular motions. Since air is a poor conductor this type of heat transfer can be virtually neglected in the atmosphere, but it is important in the ground.

(3) Convection. This occurs in fluids (including gases) which are able to circulate internally and distribute heated parts of the mass. The low viscosity of air and its consequent ease of motion makes this the chief method of atmospheric heat transfer. It should be noted that *forced convection* (mechanical turbulence) occurs due to the development of eddies as air flows over uneven surfaces, even when there is no surface heating to set up *free* (thermal) *convection*.

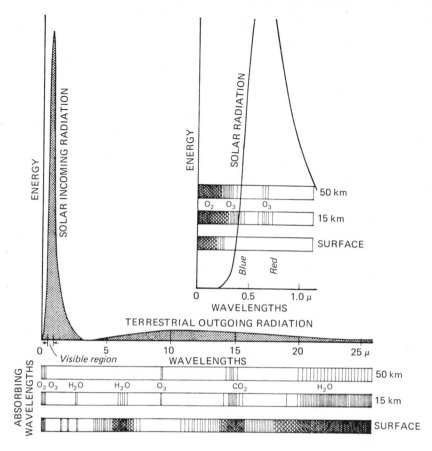

Fig. 1.8. The spectral distribution of solar and terrestrial radiation, showing the striking difference between them. The shaded strips indicate the wavelengths in which the atmosphere absorbs radiation at the surface, 15 km and 50 km. These show that much of the absorption involves the outgoing terrestrial radiation by the lowest layers of the atmosphere (from Dobson, 1963).

2. Effect of the atmosphere. Solar radiation is virtually all in the short wavelength range, less than 4 μ (Fig. 1.8). About 15% of the incoming energy is absorbed directly by ozone and water vapor. Ozone absorbs all ultraviolet radiation below 0.29 μ (2900 Å) and water vapor absorbs to a lesser extent in several narrow bands between about 0.9 μ to 2.1 μ (see Fig. 1.8). Nearly 40% is immediately reflected back into space from the atmosphere, clouds, and the earth's surface, leaving only about 60% to actually heat the earth and its atmosphere. Of this, the greater part eventually heats the atmosphere, but most of this heat is received secondhand by the atmosphere via the earth's surface, primarily in the form of sensible and latent heat. The ultimate retention of this energy by the atmosphere is of prime importance, because if it did not occur the average temperature of the earth's surface would fall by some 40°C (approximately 70°F), making most life obviously impossible. The earth itself directly absorbs 27% of the incoming short waves (together with an indirect 20% of energy reflected down or conducted from the atmosphere) and reradiates them outwards as long (infrared) waves of greater than 3 μ (Fig. 1.8). Much of this reradiated long-wave energy can be absorbed by the water vapor, carbon dioxide, and ozone in the atmosphere, the rest escaping through *radiation windows* back into outer space. Of an assumed 100% of energy available at the top of the atmosphere, only 47% is absorbed by the earth's surface. Fig. 1.9 illustrates the relative roles of the atmosphere, clouds, and the earth's surface in reflecting and absorbing solar radiation at different latitudes. (A more complete analysis of the total heat budget of the earth-atmosphere system is given in Chapter 1, Section E.)

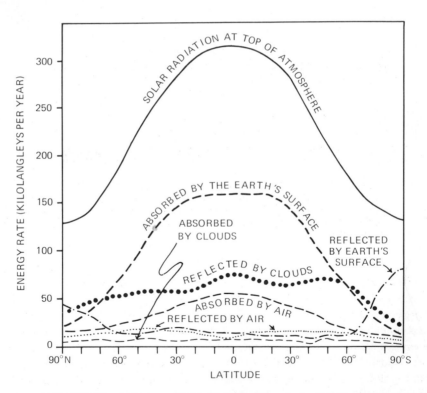

Fig. 1.9. The average annual latitudinal disposition of solar radiation (in kilolangleys; a kilolangley = 1000 cal/cm²). Of 100% radiation entering the top of the atmosphere, 23% is reflected back to space by clouds, 6% by air (plus dust and water vapor), and 7% by the earth's surface. 3% is absorbed by clouds, 14% by the air, and 47% by the earth (from Sellers, 1965).

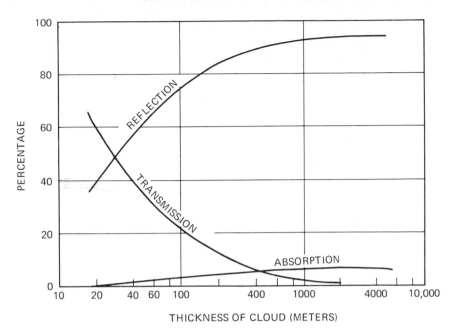

Fig. 1.10. Percentage of reflection, absorption, and transmission of solar radiation by cloud layers of different thickness (from Hewson and Longley, 1944).

3. Effect of cloud cover. Cloud cover can, if it is thick and complete enough, form a significant barrier to the penetration of insolation. How much insolation is actually reflected depends on the amount of cloud cover and thickness (Fig. 1.10). Cloud type is also important. For example, the proportion of solar radiation reflected by a complete overcast ranges from 44 to 50% for cirrostratus to 55 to 80% for stratocumulus. The total solar radiation (direct, Q, and diffuse, q) received at the surface on cloudy days is

$$Q + q = (Q + q)_0[\beta + (1 - \beta)(1 - c)],$$

where $(Q + q)_0$ = total radiation for clear skies;

   c = cloudiness (tenths);

   β = a coefficient, depending on cloud type and thickness, and the depth of the atmosphere through which the radiation must pass. For mean monthly values for the United States $\beta \simeq 0.35$, so that

$$(Q + q) \simeq (Q + q)_0[1 - 0.65c].$$

The effect of a cloud cover also operates in reverse, since it serves to retain much of the heat that would otherwise be lost from the earth by radiation throughout the day and night. This largely negative role of clouds means that their presence appreciably lessens the daily temperature range by preventing high maxima by day and low minima by night. As well as interfering with the transmission of radiation, clouds act as temporary thermal reservoirs for they absorb a certain proportion of the energy which they intercept. The effect of this absorption of solar radiation is illustrated in Figs. 1.9 and 1.10.

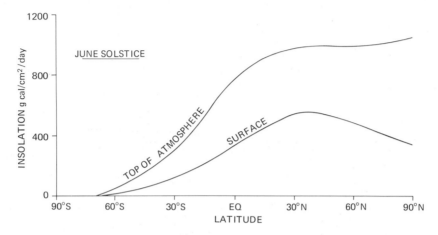

Fig. 1.11. The average receipt of insolation with latitude at the top of the atmosphere and at the earth's surface during the June solstice.

4. Effect of latitude. As Fig. 1.7 has already shown, different parts of the earth's surface receive different amounts of insolation. The time of the year is one factor controlling this, more insolation being received in summer than in winter because of the higher altitude of the sun and the longer days. Latitude is a very important control over insolation because the geographical situation of a region will determine both the duration of daylight and the distance traveled through the atmosphere by the oblique rays from the sun. However, actual calculations show the effect of the latter to be negligible in the Arctic, apparently due to the low vapor content of the air limiting the tropospheric absorption. Fig. 1.11 shows that in the upper atmosphere over the north pole there is a marked maximum of insolation at the June solstice, yet only about 30% is absorbed at the surface. This may be compared with the global average of 47% of solar radiation being absorbed at the surface. The explanation lies in the high average cloudiness over the Arctic in summer and also in the high reflectivity of the snow and ice surfaces. This example illustrates the complexity of the radiation budget and the need to take into account the interaction of several factors.

A special feature of the latitudinal receipt of insolation is that the maximum temperatures experienced at the earth's surface do not occur at the equator,

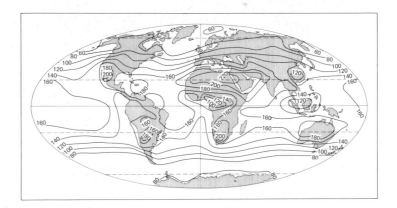

Fig. 1.12. The average annual solar radiation on a horizontal surface at ground level in kilolangleys per year (after Budyko). Maxima are found in the world's hot deserts, where as much as 80% of the solar radiation annually incident on the top of the unusually clear atmosphere reaches the ground.

as one might expect, but at the tropics. A number of factors need to be taken into account. The apparent migration of the vertical sun is relatively rapid during its passage over the equator but its rate slows down as it reaches the

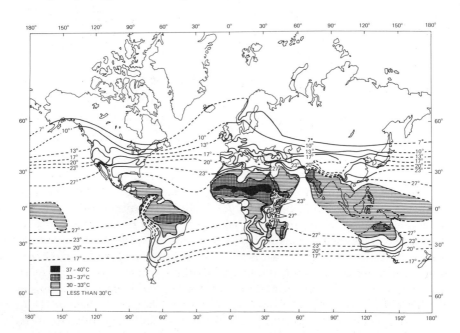

Fig. 1.13. Mean daily maximum shade air temperatures (°C) (after Ransom, 1963).

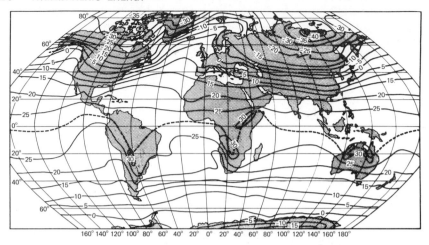

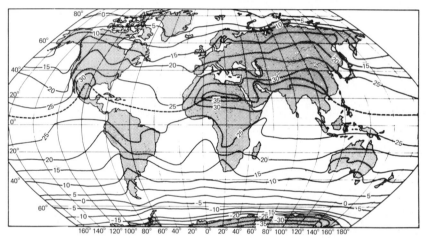

Fig. 1.14. Mean sea-level temperatures in January and July (°C). The positions of the thermal equator are approximately shown by the dashed line.

tropics. Between 6°N and 6°S the sun's rays remain almost vertically over-head for only 30 days during each of the spring and autumn equinoxes, allowing little time for any large build-up of surface heat and high tempera-tures. On the other hand, between 17.5° and 23.5° latitude the sun's rays shine down almost vertically for 86 consecutive days during the period of the solstice. This longer sustained period, combined with the fact that the tropics experience longer days than at the equator, makes the maximum zones of heating occur nearer the tropics than the equator. In the northern hemisphere this poleward displacement of the zone of maximum heating is emphasized by the effect of *continentality* (see Chapter 1, Section D.5), while low cloudiness associated with the subtropical high-pressure belts is an additional factor. The clear skies are particularly effective in allowing

large annual receipts of solar radiation in these areas (Fig. 1.11). The net
result of these influences is shown by Fig. 1.12 in terms of the average annual
solar radiation on a horizontal surface at ground level, and by Fig. 1.13 in
terms of the average daily maximum shade temperatures. Over the continents the highest values occur at about 23°N and 10°–15°S. In consequence the mean annual _thermal equator_ (that is, the zone of maximum
temperature) is located at about 5°N. Nevertheless the mean surface temperatures of the earth are very broadly related to latitude (Fig. 1.14).

5. Effect of land and sea. Another important control on the effect of incoming solar radiation stems from the different ways in which land and sea
are able to profit from it. Whereas water has a tendency to store the heat it
receives, land, in contrast, quickly returns it to the atmosphere. There are
several reasons for this.

A large proportion of insolation is reflected back into the atmosphere
without heating the earth's surface at all. The proportion of incident radiation that is reflected is termed the _albedo_, or reflection coefficient (α).
It depends upon the type of surface. For land surfaces the albedo is generally
between 8% and 40% of the incoming radiation. The figure for forests is
about 9 to 18% according to the type of tree and density of foliage (see
Chapter 7, Section B), for grass approximately 25%, for cities 14 to 18%,
and desert sand 30 to 40%. Fresh, flat snow may reflect as much as 85% of
solar radiation, whereas a sea surface reflects very little unless the angle of
incidence of the sun's rays is small. The albedo for a calm water surface is
only 2 to 3% for a solar elevation angle exceeding 60°, but is more than 50%
when the angle is 15°.

The solar radiation absorbed at the surface is determined from measurements of incident radiation and albedo. It may be expressed as

$$(Q + q)(1 - \alpha)$$

where the albedo is in hundredths. A snow surface will absorb only about
15% of the incident radiation, whereas for the sea the figure generally
exceeds 90%. The ability of the sea to absorb the heat received also depends
upon its transparency. As much as 10% of the radiation penetrates as far
down as 9 meters (30 ft). Fig. 1.15 provides some indication of how much
energy is absorbed by the sea at different depths. However, the heat absorbed
by the sea is carried down to considerable depths by the turbulent mixing of
water masses by the action of waves and currents. Fig. 1.16, for example,
illustrates the warming of the North Sea down to about 40 meters in summer.
In completely _still_ water the annual heat penetration would only be apparent
down to about 3–4 meters.

A measure of the difference between the subsurfaces of land and sea is
given in Fig. 1.17, which shows ground temperatures at Kaliningrad (Königs-

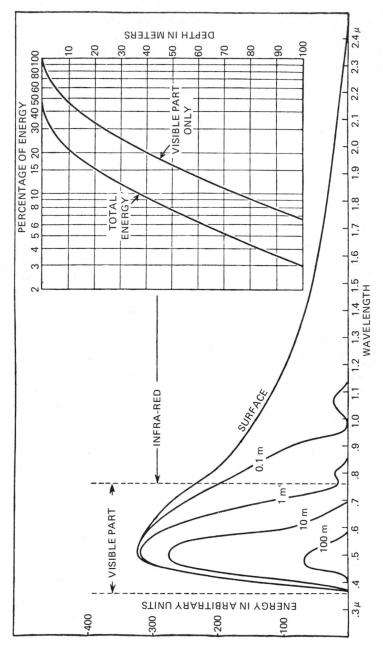

Fig. 1.15. Schematic representation of the energy spectrum of the sun's radiation (in arbitrary units) which penetrates the sea surface to depths of 0.1, 1, 10 and 100 meters. This illustrates the absorption of infra-red radiation by water, and also shows the depths to which visible (light) radiation penetrates (from Sverdrup, 1945).

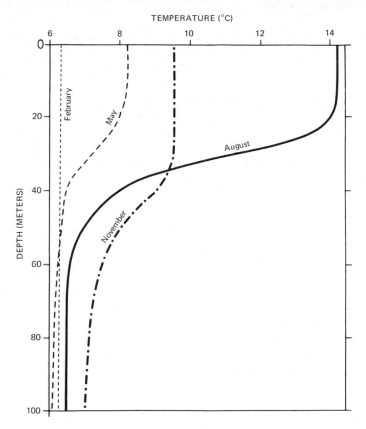

Fig. 1.16. *Mean temperatures of the upper 100 meters of the North Sea for February, May, August and November (from Lumb, 1961) (Crown Copyright Reserved).*

Fig. 1.17. *Annual variation of temperature at different depths in soil at Kaliningrad (left) and in the water of the Bay Biscay (at approximately 47°N, 12°W) (right), illustrating the relatively deep penetration of solar energy into the oceans as distinct from that into land surfaces. The right-hand figure shows the temperature deviations from the annual mean for each depth (from Geiger, 1965 and Sverdrup, 1945).*

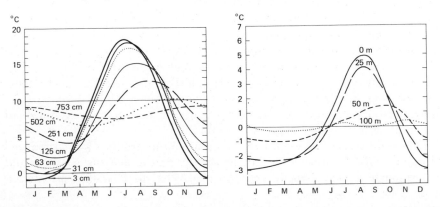

berg) and sea temperature deviations from the annual mean at various depths in the Bay of Biscay. Heat transmission in the soil is carried out almost wholly by conduction and the degree of conductivity varies with the moisture content and porosity of each particular soil.

Air is an extremely poor conductor and for this reason a loose, sandy soil surface heats up rapidly by day, as the heat is not conducted away. Increased soil moisture tends to raise the conductivity by filling the soil pores, but too much moisture increases the soil's heat capacity, thereby reducing the temperature response. The relative depths over which the annual and diurnal temperature variations are effective in wet and dry soils are roughly as follows:

	Diurnal variation	Annual variation
Wet soil	0.5 m	9 m
Dry sand	0.2 m	3 m

However, the *actual* temperature change is greater in dry soils. For example, the following values of diurnal temperature range have been observed during clear summer days at Sapporo, Japan:

	Sand	Loam	Peat	Clay
Surface	40°C	33°C	23°C	21°C
5 cm	20	19	14	14
15 cm	7	6	2	4

The different heating qualities of land and water are also partly to be accounted for by their different *specific heats*. The specific heat of a substance can be represented by the number of thermal units (calories) required to raise a unit mass (gram) of it through one degree (Celsius). The specific heat of water is much greater than for most other common substances, and thus water must absorb five times as much heat energy to raise its temperature by the same amount as a comparable mass of dry soil. If unit volumes of water and soil are considered the heat capacity of the water exceeds that of the soil approximately twofold. When this water is cooled the situation is reversed, for then a large quantity of heat is released. A meter-thick layer of sea water being cooled by as little as 0.1°C will release enough heat to raise the temperature of approximately a 30-meter-thick air layer by 10°C (18°F). In this way the oceans act as a very effective reservoir for much of the world's heat. Similarly evaporation of sea water causes a large heat expenditure because a great amount of energy is needed to evaporate even a small quantity of water (see Chapter 2, Section A).

These differences between land and sea help to produce what is termed continentality. Continentality implies, firstly, that a land surface heats and cools much quicker than that of an ocean. Over the land the lag between

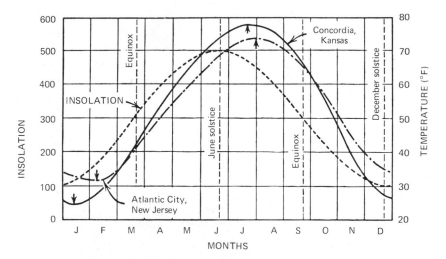

Fig. 1.18. Annual distributions of temperature at maritime (Atlantic City) and continental (Concordia, Kansas) mid-latitude locations. Representative insolation receipts are shown by the dashed line (in cal/cm²/day). The respective time lags of temperature behind the curve of insolation are indicated by the arrows (data from Trewartha. From Strahler, 1960).

Fig. 1.19. Mean annual temperature regimes for Poona (Monsoon), Brazzaville (Equatorial), Stornoway (Temperate Maritime) and Winnipeg (Temperate Continental).

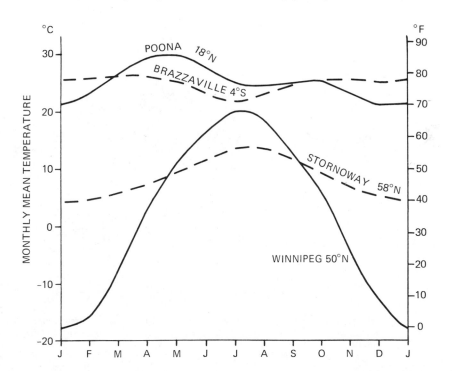

maximum and minimum periods of insolation and the maximum and minimum surface temperatures is only one month, but over the ocean and at coastal stations the lag is as much as two months. (Fig. 1.18). Secondly, the annual and diurnal ranges of temperature are greater in continental than in coastal locations. Fig. 1.19 illustrates the annual variation of temperature at Winnipeg and Stornoway, while Fig. 1.23C shows the diurnal ranges experienced in continental and maritime areas. This is discussed more fully below. The third effect of continentality results from the global distribution of the land masses. The small sea area of the northern hemisphere causes the north-

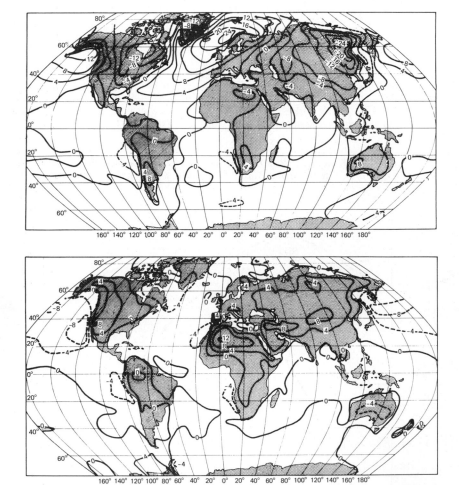

Figl 1.20. World temperature anamolies (or the difference between recorded temperatures (°C) and the mean for that latitude) for January and July. Solid lines indicate positive, and dashed lines negative, anomalies.

ern hemisphere summer to be warmer but its winters colder on the average than those of the southern hemisphere (summer, 22.4°C (72.3°F) versus 17.1°C (62.7°F); winter, 8.1°C (46.5°F) versus 9.7°C (49.5°F)). Heat storage in the oceans causes them to be warmer in winter and cooler in summer than land in the same latitude, although ocean currents give rise to some local departures from this rule. The distribution of temperature anomalies for the latitude in January and July (Fig. 1.20) illustrates the significance of continentality and also the influence of the warm drift currents in the North Atlantic and the North Pacific in winter (compare Fig. 3.29).

6. Effect of elevation and aspect. When we come down to the local scale even differences in the elevation of the land and its aspect (that is, the direction which the surface faces) will strikingly control the amount of insolation received. Obviously some slopes are more exposed to the sun than others, while really high elevations which have a much smaller mass of air above them (see Fig. 1.4) receive considerably more insolation under clear skies than locations near sea level. On average in middle latitudes the intensity of incident solar radiation increases by 5–15% for each 1000 meters increase in elevation in the lower troposphere. The difference between sites at 200 and 3000 meters in the Alps, for instance, can amount to 140 cal/cm²/day in cloudless summer conditions. However, there is also a correspondingly greater net loss of terrestrial radiation at higher elevations because the low density of the overlying air results in a smaller fraction of the outgoing radiation being absorbed. The overall effect is invariably complicated by the greater cloudiness associated with most mountain ranges and it is therefore impossible to generalize from the limited data at present available.

The effect of aspect and slope angle on insolation receipts is illustrated in Fig. 1.21. The radiation intensity on a sloping surface (I_s) is

$$I_s = I_0 \cos \gamma,$$

where γ = the angle between the solar beam and a beam normal to the sloping surface. Relief may also affect the quantity of insolation and the duration of direct sunlight when a mountain barrier screens the sun from valley floors and sides at certain times a day. In many alpine valleys settlement and cultivation are noticeably concentrated on southward-facing slopes (the adret or sunny side), whereas northward slopes (ubac or shaded side) remain forested.

E. HEAT BUDGET OF THE EARTH

In the preceding section we described how the earth is heated by the sun's rays and listed the main factors that can cause variations in the insolation receipt of different regions. Forgetting for a moment these variations, it is

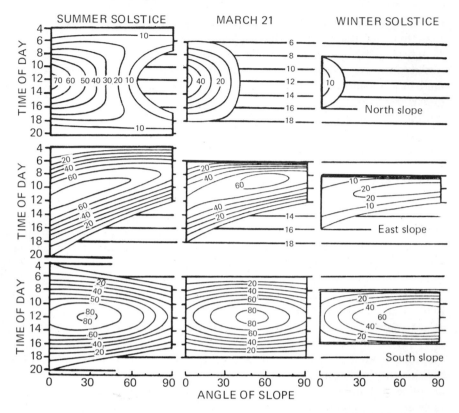

Fig. 1.21. The amount of insolation received under cloudless skies at Trier, West Germany (50°N) for north-, east- and south-facing slopes of all inclinations for the solstices and for 21 March. The units are cal/cm²/hour (after calculations by W. Kaempfert. From Geiger, 1965).

worthwhile summarizing the vertical pattern of heat transfers for the earth-atmosphere system.

The radiation from the sun reaches the earth predominantly in the form of short waves and leaves as long waves (see Fig. 1.8). Of all the incoming radiation, regarded for convenience as 100 units, 2 units are absorbed in the stratosphere, by ozone mainly, and 15 units (A) are absorbed in the tropo-sphere by ozone, water vapor, and water droplets in clouds; 23 units are reflected back into space from clouds (B) and 7 units from the earth's surface (E), while a further 6 units are scattered upwards by air molecules, water drops, and dust particles (C). The remaining 47 units (D) reach the earth either directly (31 units) or as diffuse radiation transmitted via clouds or by downward scattering (16 units). The scattering effect of air molecules on the visible wavelengths of radiation (blue light = 0.4 μ, red = 0.7 μ) is greatest at short wavelengths, and hence the sky light appears blue in color.

The pattern of outgoing terrestrial radiation is quite different (Fig. 1.22). The black-body radiation, assuming a mean surface temperature of 288°K (15°C), is equivalent to 98 units of long-wave (infrared) radiation—but this does not imply that the surface undergoes a net loss of radiation, as the figures for total balance show. The great majority of this is absorbed in the atmosphere (G) chiefly by carbon dioxide, water vapor, and cloud droplets, although 7 units (J) are lost through radiation windows—bands in the wavelength spectrum where little or no absorption takes place. The troposphere reradiates 78 units back to the surface of the earth (H) and 57 units to space (I).

It is worth emphasizing that long-wave radiation is not merely terrestrial. The atmosphere itself radiates to space, and clouds are particularly important since they act as black bodies.

These radiation transfers can be expressed symbolically:

$$R_n = (Q + q)(1 - \alpha) + L_n,$$

where R_n = net radiation, $(Q + q)$ = total incident solar radiation, α = albedo, and L_n = net longwave radiation $(H - F)$. At the surface R_n = 27 units. This surplus is conveyed to the atmosphere by the turbulent transfer of sensible heat (5 units, L) and latent heat (22 units, K).

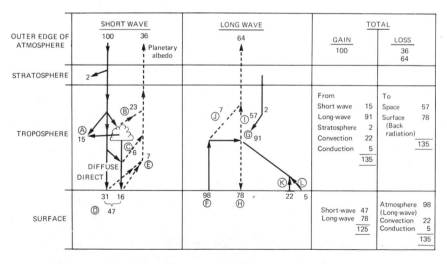

Fig. 1.22. The balance of the atmospheric energy budget (data after Budyko and others). The transfers are explained in the text. Solid lines indicate energy gains by the atmosphere and surface in the left-hand diagram and by the troposphere in the right-hand diagram. The exchanges are referred to 100 units of incoming solar radiation at the top of the atmosphere (equal to 0.5 cal/cm²/min or 263 kcal/cm²/yr).

$$R_n = LE + H,$$

where H = sensible heat transfer and LE = latent heat transfer. There is also a flux of heat into the ground (Chapter 1, Section D.5), but for annual averages this is approximately zero.

The total balances at the surface (± 125 units) and in the troposphere (± 135 units) are illustrated in Fig. 1.22. The energy balance for the earth-atmosphere system is estimated to be ± 170 Kcal/cm²/year (± 69 units). At the outer atmosphere the balance must be ± 100 units, equivalent to ± 263 Kcal/cm²/year (or 0.5 cal/cm²/min). This figure represents the incident solar radiation averaged over the globe, which is

$$\text{Solar constant} \times \frac{\pi r^2}{4\pi r^2},$$

where r = the radius of the earth and $4\pi r^2$ is the surface area of a sphere.

Present estimates of the global heat budget are still rather crude, although satellites are now providing a "top-view" of the radiation exchanges (see Plates 1 and 2). Such measurements should help to resolve some of the present uncertainties.

The annual and diurnal variations of temperature are directly related to the local radiation budget. The sum of the net upward and downward radiation transfers at the surface is, in most latitudes, positive between about one hour after sunrise to a few hours before sunset (Fig. 1.23A). At the equinox in middle latitudes the diurnal temperature maximum occurs between 2 to 3 PM when the net radiation is zero. At this hour the long-wave radiation loss is a maximum. The minimum temperature occurs in the early morning when the net radiation is again zero, and at this hour the long-wave emission is much less than by day due to the cold surface. The annual pattern of the radiation budget and temperature regime is directly analogous to the diurnal one (Fig. 1.23B).

There are marked latitudinal variations in the diurnal and annual ranges of temperature. Broadly, the annual range is a maximum in higher latitudes, with extreme values about 65°N related to the effects of continentality in Asia and North America. The diurnal range reaches a maximum at the tropics over land areas, but in the equatorial zone the diurnal variation of heating and cooling exceeds the annual one (Fig. 1.23C). This is of course related to the small seasonal change in solar elevation angle at the equator.

F. ATMOSPHERIC ENERGY AND HORIZONTAL HEAT TRANSPORT

So far, we have described the gases and other constituents which make up our atmosphere, and have given some account of the earth's heat budget. We have already referred to two forms of energy—internal (or heat) energy due to the motion of individual air molecules and latent energy which is

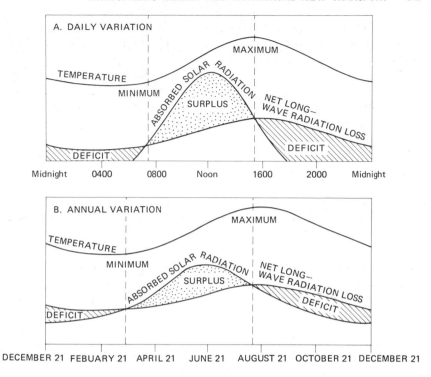

Fig. 1.23. *Curves showing the daily (A) and annual (B) marches of temperature charac-*
teristic of most stations. The lag between the maximum incoming solar radiation and
maximum temperature occurs because the temperature continues to rise as long as incoming
energy exceeds outgoing (adapted from Miller, 1966). C. The latitudinal distribution of
the annual and diurnal temperature range in continental and maritime areas (from Paffen,

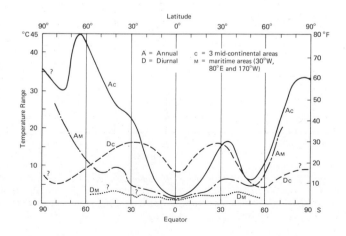

released by condensation of water vapor. Two other forms of energy are important—geopotential energy due to gravity and height above the surface and kinetic energy associated with air motion.

Geopotential and internal energy are interrelated since the addition of heat to an air column not only increases its internal energy, but adds to its geopotential as a result of the vertical expansion of the air column. In a column extending to the top of the atmosphere the geopotential is approximately 40% of the internal energy. These two are therefore usually considered together and termed the total potential energy (PE). For the whole atmosphere

$$\text{potential energy} \simeq 10^{24} \text{ joules}[5] \quad (23.9 \times 10^{22} \text{ calories})$$
$$\text{kinetic energy} \simeq 10^{20} \text{ joules.}$$

In a later section (Chapter 3, Section E) we shall see how energy is transferred from one form to another, but here we need only be concerned with heat energy. It is apparent that the receipt of heat energy is very unequal geographically and that this must lead to great lateral transfers of energy across the surface of the earth. Much present-day meteorological research is focused

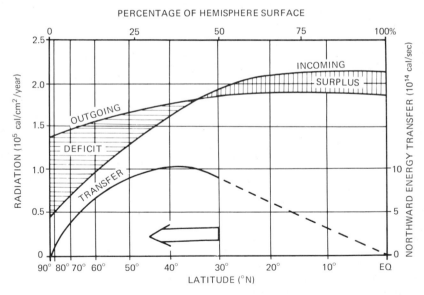

Fig. 1.24. A meridional illustration of the balance between incoming solar radiation and outgoing radiation from the earth (data from Houghton. After Newell, 1964), in which the zones of permanent surplus and deficit are maintained in equilibrium by a poleward energy transfer (after Gabites).

---

[5] See Appendix 3.

on these transfers, since undoubtedly they give rise, at least indirectly, to the observed patterns of global weather and climate.

The amounts of energy received at different latitudes vary substantially, the equator on the average receiving 2.5 times as much annual solar energy as the poles. Clearly if this process was not modified in some way the variations in receipt would cause a massive accumulation of heat within the tropics (associated with gradual increases of temperature) and a corresponding deficiency at the poles. Yet this does not seem to happen, and the earth as a whole is roughly in a state of thermal equilibrium in so far as no one region is obviously gaining heat at the expense of another. Some authors believe the Ice Ages to have been an exception to this rule. One explanation of this equilibrium could be that for each region of the world there is an equalization between the amounts of incoming and outgoing radiation. Observation shows that this is not so (Fig. 1.24), however, for, whereas incoming radiation varies appreciably with changes in latitude, being highest at the equator and declining to a minimum at the poles, outgoing radiation seems to maintain a more even latitudinal distribution. Some other explanation therefore becomes necessary.

1. The horizontal transport of heat. If the net radiation for the whole earth-atmosphere system is calculated, it is found that there is a positive budget between 35°S and 40°N as shown in Fig. 1.25A. As the tropics do not get progressively hotter or the high latitudes colder, a redistribution of world heat energy must constantly occur, taking the form of a continuous movement of energy from the tropics to the poles. In this way the tropics shed their excess heat and the poles are not allowed to reach extremes of cold. If there were no meridional interchange of heat, a radiation balance at each latitude would only be achieved if the equator were 14°C warmer and the north pole 25°C colder than now. This poleward heat transport takes place within the atmosphere and oceans, and it is estimated that the former accounts for approximately 80% of the required total. The horizontal transport (*advection* of heat) occurs in the form of both latent heat (that is, water vapor which subsequently condenses) and sensible heat (that is, warm air masses) (Fig. 1.25B). It varies in intensity according to the latitude and the season. Fig. 1.25B shows the mean annual pattern of energy transfer by the three mechanisms. The latitudinal zone of maximum total transfer rate is found between latitudes 35° and 45° in both hemispheres, although the patterns for the individual components are quite different from one another. The latent heat transport, which occurs almost wholly in the lowest two or three kilometers, reflects the global wind belts on either side of the subtropical high-pressure zones (see Chapter 3, Section D). The more important meridional transfer of sensible heat has a double maximum not only latitudinally but also in the vertical plane, where there are maxima near the surface and at about 200 mb. The high-level transport is particularly significant over the subtropics,

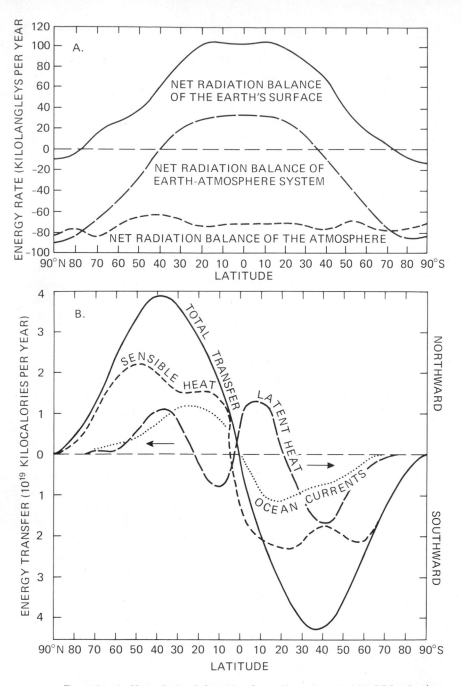

Fig. 1.25. A. Net radiation balance for the earth's surface of +72 kilolangleys/year (incoming solar radiation of 124 kly/yr minus outgoing long-wave energy to the atmosphere of 52 kly/yr); for the atmosphere of −72 kilolangleys/year (incoming solar radiation of 45 kly/yr minus outgoing long-wave energy to space of 117 kly/yr); and for the whole earth-atmosphere system of zero (from Sellers, 1965). B. The average annual latitudinal distribution of the components of the poleward energy transfer (in 10^19 kilocalories) in the earth-atmosphere system (from Sellers, 1965).

whereas the primary latitudinal maximum about 50° to 60°N is related to the travelling low-pressure systems of the westerlies.

The intensity of the poleward energy flow is closely related to the meridional (that is, north–south) temperature gradient. In winter this temperature gradient is at a maximum and in consequence the hemispheric air circulation is most intense. The nature of the complex transport mechanisms will be discussed in Chapter 3, Section E.

As shown in Fig. 1.25B, ocean currents account for a significant proportion of the poleward heat transfer in low latitudes. The Gulf Stream and Kuro Shio currents are particularly important. As a result of this factor, the energy budget equation for an ocean area must be expressed as

$$R_n = LE + H + G + \Delta F,$$

where ΔF = horizontal advection of heat by currents and G = the heat transferred into or out of storage in the water. The latter is more or less zero for annual averages.

2. Spatial pattern of the heat budget components. The mean latitudinal values of the heat budget components discussed above conceal great spatial variations. Fig. 1.26 shows the global distribution of the annual net radiation. Broadly, its magnitude decreases poleward from about 25° latitude, although as a result of the considerable absorption of solar radiation by the sea, the net radiation is greater over the oceans—exceeding 120 Kcal/cm² in latitudes 15–20°—than over land areas, where it is about 60–80 Kcal/cm² in the same latitudes. Net radiation is also rather lower in arid continental areas

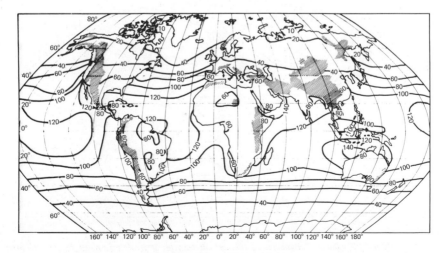

Fig. 1.26. Global distribution of the annual net radiation, in Kcal/cm² (after Budyko).

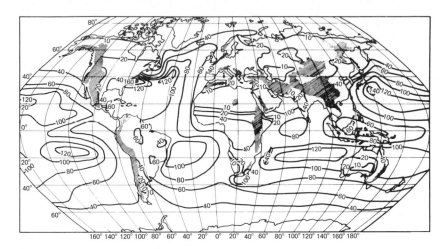

Fig. 1.27. Global distribution of the vertical transfer of latent heat, in Kcal/cm² (after Budyko).

than in humid ones, because in spite of the increased insolation receipts under clear skies there is at the same time greater net loss of terrestrial radiation.

Figs. 1.27 and 1.28 show the vertical transfers of latent and sensible heat to the atmosphere. Both maps show that the fluxes are distributed very differently over land and sea. Heat expenditure for evaporation is at a maximum in tropical and subtropical ocean areas, where it exceeds 120 Kcal/cm²/year. It is less near the equator where wind speeds are somewhat lower and the air has a vapor pressure close to the saturation value (see Chapter 2, Section A). It is clear from Fig. 1.27 that the major warm cur-

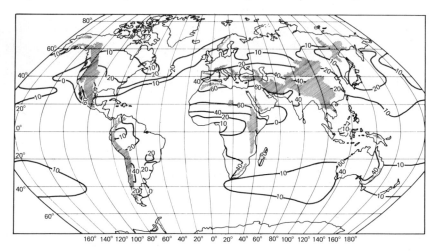

Fig. 1.28. Global distribution of the vertical transfer of sensible heat, in Kcal/cm² (after Budyko).

rents considerably augment the evaporation rate. On land the latent heat transfer is greatest in hot, humid regions. It is least in arid areas due to the low precipitation and in high latitudes where there is little available energy.

The largest exchange of sensible heat occurs in the tropical deserts where more than 60 Kcal/cm²/year is transferred to the atmosphere. In contrast to latent heat, the sensible heat flux is generally small over the oceans, only reaching 20–30 Kcal/cm² in areas of warm currents. Indeed, negative values occur (transfer *to* the ocean) where warm continental air masses move offshore over cold currents.

G. THE LAYERING OF THE ATMOSPHERE

The atmosphere can be divided conveniently into a number of rather well-marked horizontal layers, mainly on the basis of temperature. The evidence for this structure comes from regular RAWINSONDE (radar wind-sounding) balloons, radio-wave investigations, and, more recently, from rocket and satellite flights. Broadly, the pattern (Fig. 1.29) consists of three relatively warm layers (near the surface; between 50 and 60 km; and above about 120 km), separated by two relatively cold layers (between 10 and 30 km; and about 80 km). Mean January and July temperature sections illustrate the considerable latitudinal variations and seasonal trends that complicate the scheme (Fig. 1.30).

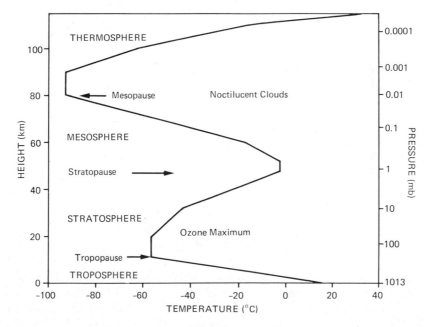

Fig. 1.29. The generalized vertical distribution of temperature and pressure up to about 110 km. Note particularly the tropopause and the zone of maximum ozone concentration with the warm layer above it (based on data in Valley, 1965).

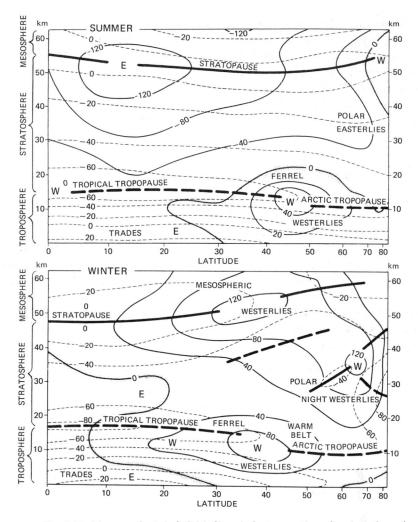

Fig. 1.30. Mean zonal winds (solid isolines, in knots; negative values from the east) and temperatures (in degrees Celsius, dashed isolines), showing the broken tropopause near the Ferrel jet stream (after Boville. From Hare, 1962).

1. Troposphere. The lowest layer of the atmosphere is called the troposphere. It is the zone where weather phenomena and atmospheric turbulence are most marked, and contains 75% of the total molecular or gaseous mass of the atmosphere and virtually all the water vapor and aerosols. Throughout this layer there is a general decrease of temperature with height at a mean rate of about 6.5°C/km (or 3.6°F/1000 ft), and the whole zone is capped in most places by a temperature inversion level (that is, a layer of relatively warm air above a colder one) and in others by a zone which is isothermal with height. The troposphere thus remains to a large extent self-contained because the inversion acts as a lid which effectively limits convection (see Chapter 2, Section G). This inversion level or weather ceiling is called the *tropopause*.[6] Its height is not constant either in space or time. It seems that the height of the tropopause at any point is correlated with sea-level temperature and pressure, which are in turn related to the factors of latitude, season, and daily changes in surface pressure. There are marked variations in the altitude of the tropopause as between different latitudes (Fig. 1.30), it having an elevation of about 16 km (10 miles) at the equator where there is great heating and convective turbulence and only 8 km (5 miles) at the poles.

The meridional temperature gradients in the troposphere in summer and winter are roughly parallel, as are the tropopauses (Fig. 1.30), and the strong lower mid-latitude temperature gradient in the troposphere is reflected in the tropopause breaks (see also Fig. 3.14). In these zones important interchanges can occur between the troposphere and stratosphere, or vice versa. Traces of water vapor probably penetrate into the stratosphere by this means, while dry, ozone-rich stratospheric air may be brought down into the mid-latitude troposphere. For example, above-average concentrations of ozone are observed in the rear of mid-latitude low-pressure systems where the tropopause elevation tends to be low. Both facts are probably the result of stratospheric subsidence, which warms the lower stratosphere and causes downward transfer of the ozone.

2. Stratosphere. The second major atmospheric layer is the stratosphere, which extends upwards from the tropopause to about 50 km (30 miles). Although the stratosphere contains much of the total atmospheric ozone (it reaches a peak density at approximately 22 km), the maximum temperatures associated with the absorption of the sun's ultraviolet radiation by ozone do not occur until the more exposed upper levels are reached at the *stratopause*, where temperatures may exceed 0°C (Fig. 1.30). Temperatures increase fairly generally with height in summer, with the coldest air at the equatorial tropopause. In winter the structure is more complex with very low temperatures, averaging −80°C at the equatorial tropopause which is highest at this season. Similar low temperatures are found in the middle stratosphere at high latitudes, whereas over 50°–60°N there is a marked warm region with nearly isothermal conditions at about −45°C to −50°C. Marked seasonal changes of temperature affect the stratosphere. The cold polar night winter stratosphere undergoes dramatic *sudden warmings* associated with subsidence due to circulation changes in late winter or early spring when temperatures at about 25 km may jump from −80°C to −40°C over a two-day period. The autumn cooling is a more gradual process. In the

---

[6] The official definition is the lowest level at which the lapse rate decreases to less than, or equal to, 2°C/km (provided that the average lapse rate of the 2-km layer above does not exceed 2°C/km).

tropical stratosphere recent investigations have revealed a biannual wind regime, with easterlies in the layer 18 to 30 km for 12 to 13 months, followed by westerlies for a similar period. The reversal begins first at high levels and takes approximately 12 months to descend from 30 to 18 km (10 to 60 mb). The reasons for this pattern are not yet clear, though again there may be a link with the distribution and transport of ozone.

How far these events in the stratosphere are linked with temperature and circulation changes in the troposphere is a major topic of current meteorological research. Any interactions that do exist, however, are likely to be complex, otherwise they would already have become evident.

3. The upper atmosphere. (a) Mesosphere.

Above the warm stratopause temperatures decrease to a minimum of about −90°C (183°K) around 80 km. This layer is commonly termed the mesosphere, although it must be noted that as yet there is no universal acceptance of terminology for the upper atmospheric layers. Indeed, some authors refer to the layer between 20 and 80 km as the mesosphere. Above 80 km temperatures again begin rising with height and this inversion is referred to as the mesopause. It is in this region that noctilucent clouds are observed over high latitudes in summer. Their presence appears to be due to meteoric dust particles which act as nuclei for ice crystals when traces of water vapor are carried upwards by high-level convection caused by the vertical decrease of temperature in the mesosphere.

Pressure is very low in the mesosphere, decreasing from about 1 mb at 50 km to 0.01 mb at 80 km.

(b) Thermosphere.

Above the mesopause atmospheric densities are extremely low, although the tenuous atmosphere still effects drag on space vehicles above 250 km. The lower portion of the thermosphere is composed mainly of nitrogen (N_2) and oxygen in molecular (O_2) and atomic (O) forms, whereas above 200 km atomic oxygen predominates over nitrogen (N_2 and N). Temperatures rise with height, owing to the absorption of ultraviolet radiation by atomic oxygen, probably approaching 1200°K at 350 km, but these temperatures are essentially theoretical. For example, artificial satellites do not acquire such temperatures because of the rarefied air. Ultraviolet radiation from the sun and high-energy particles from outer space (cosmic rays) enter the atmosphere above 100 km at high velocity and cause *ionization*, or electrical charging, by separating negatively charged electrons from oxygen atoms and nitrogen molecules. The Aurora Borealis and Aurora Australis are produced by the penetration of ionizing particles through the atmosphere from about 300 to 80 km, particularly in zones about 20°–25° latitude from the earth's magnetic poles. On occasion, however, the aurorae may appear at heights up to 1000 km, demonstrating the immense extension of a rarefied atmosphere. The term *ionosphere* is commonly applied to the layers above 80 km, although sometimes it is used only for the region of high electron density between about 100 and 300 km. In view of these different designations it seems preferable to avoid confusion by using the terminology adopted here.

(c) Exosphere and magnetosphere. The base of the exosphere is between about 500 and 750 km. Here neutral atomic oxygen, ionized oxygen, and hydrogen atoms form the tenuous atmosphere and the gas laws (see Chapter 1, Section B) cease to be valid. Gas particles, especially helium which has low atomic weight, can escape into space since the chance of molecular collisions to deflect them downwards becomes less with increasing height. An equivalent supply of helium is probably produced by the action of cosmic radiation on nitrogen and from the slow but steady breakdown of radioactive elements in the earth's crust. Neutral particles are still predominant, but beyond about 2000 km, in the magnetosphere, there are only electrons (negatively charged) and protons (positively charged) and the earth's magnetic field becomes more important than gravity in their distribution. The charged particles are concentrated in two zones at about 4000 km and 20,000 km, known as the Van Allen radiation belts. Detailed investigation of these regions has been made possible since 1958 by satellites, but study of this outermost fringe lies in the field of magnetohydrodynamics. Nevertheless disturbances of these upper regions by streams of charged hydrogen particles from the sun (the solar wind) may eventually prove to have meteorological significance at lower levels. At a height of 80,000 km or so, the earth's atmosphere probably merges into that of the sun, but even the appropriate definitions of atmosphere, wind, and temperature are uncertain in these regions.

H. VARIATION OF TEMPERATURE WITH HEIGHT

The last section described the gross characteristics of the vertical temperature profile in the atmosphere, but it is now necessary to examine in more detail some of the features of the temperature gradient at low levels.

Vertical temperature gradients are determined in part by energy transfers and in part by vertical motion of the air. The various factors interact in a highly complex manner. The energy terms are the release of latent heat by condensation (Chapter 2, Section E), radiational cooling of the air, and sensible heat transfer from the ground. Horizontal temperature advection may also be important. Vertical motion is dependent on the type of pressure system. High-pressure areas are generally associated with descent and warming of deep layers of air, hence decreasing the temperature gradient and frequently causing temperature inversions in the lower troposphere. In contrast, low-pressure systems are associated with rising air which cools upon expansion and increases the vertical temperature gradient. This is only part of the story since moisture is an additional complicating factor (see Chapter 2, Section F). It remains true, however, that the middle and upper troposphere is relatively cold above a surface low-pressure area, leading to a steeper temperature gradient.

The overall vertical decrease of temperature, or *lapse rate*, in the troposphere is, as has been stated, about 6.5°C/km. However, this is by no means constant with height, season, or location. Average global values calculated by C. E. P. Brooks for July show increasing lapse rate with height: 5°C/km in the lowest 2 km, 6°C/km between 4 and 6 km, and 7°C/km between 6 and 8 km. Winter values are generally smaller and in continental areas, such as central Canada or eastern Siberia, may even be negative (that is, temperatures increase with height in the lower layer) as a result of excessive radiational cooling over a snow surface (Fig. 1.31). A similar effect is common in mountain basins where dense cold air flows down the slopes to accumulate in the valley bottoms. On such occasions the mountain tops may be many degrees warmer than the valley floor below (see Chapter 3, Section B.2). For this reason the adjustment of average temperatures of upland stations to mean sea level may produce misleading results. Observations in Colorado at Pike's Peak (14,111 ft, or 4,301 meters) and Colorado Springs (6,098 ft, or 1,859 meters) show the mean lapse rate to be 4.1°C/km (2.3°F/1000 ft) in winter and 6.2°C/km (3.4°F/1000 ft) in summer! It should be noted that such topographic lapse rates sometimes bear little relation to free-air lapse rates, and the two must be carefully distinguished.

Table 1.3 summarizes the seasonal characteristics of lapse rates in six major climatic zones and examples of five of these are illustrated in Fig. 1.31. The seasonal regime is very pronounced in continental areas with cold winters whereas inversions persist for much of the year in the Arctic. In winter the Arctic inversion is due to intense radiational cooling but in summer it is the result of the surface cooling of advected warmer air. The winter lapse rate is only greater than the summer one in Mediterranean climates. In these regions there is more likelihood of rising air associated with low-

Table 1.3: Temperature lapse rates in the lowest 1000–1500 meters
(*after Lautensach and Bögel*).

Climate	Season of maximum	Rate °C/km	Season of minimum	Rate °C/km
Tropical rainy	Dry season	>5	Rainy season	>4.5
Tropical and subtropical deserts	Summer	>8	Winter	>5
Mediterranean	Winter	>5	Summer	<5
Mid-latitudes (cold winter)	Summer	>6	Winter	0 − 5
Boreal continental	Summer	>5	Winter	<0
Arctic	Summer	≤0	Winter	<0

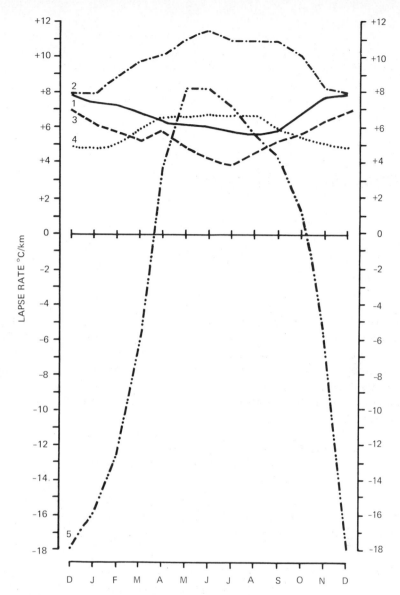

Fig. 1.31. The annual variation of lapse rate in five climatic zones (from Hastenrath, 1968).
1. Tropical rainy climate (Togo).
2. Tropical desert (Arizona).
3. Mediterranean (Sicily).
4. Mid-latitude, cold winter climate (North Germany).
5. Boreal continental (Eastern Siberia).

pressure areas in winter. In contrast, subsidence is predominant in the desert zones in winter. The tropical and subtropical deserts have very steep lapse rates in summer when there is considerable heat transfer from the surface and generally ascending motion.

2

Atmospheric Moisture

Terrestrial moisture is in a constant state of transformation, termed the hydrologic cycle, in which the three most important stages are evaporation, condensation, and precipitation. Fig. 2.1 indicates the relative average annual amounts of water involved in each phase of the cycle. It shows that the atmosphere holds only a very small amount of water although the exchanges with the land and oceans are very considerable. This is further emphasized by the following table:

Table 2.1: Mean water content of the atmosphere
(in cm of rainfall equivalent). (After Sutcliffe, 1956.)

	Northern Hemisphere	Southern Hemisphere	World
January	1.9 (0.8 in.)	2.5 (1.0 in.)	2.2 (0.9 in.)
July	3.4 (1.3 in.)	2.0 (0.8 in.)	2.7 (1.1 in.)

The average storage of water in the atmosphere (about 2.5 cm, or 1 in.) is only sufficient for some 10 days' supply of rainfall over the earth as a whole. However, intense (horizontal) influx of moisture into the air over a given region makes possible short-term rainfall totals greatly in excess of 1 in. The phenomenal record total of 187 cm (73.6 in.) fell on the island of Réunion, off Madagascar, during 24 hours in March 1952, and much greater intensities have been observed over shorter periods (see Chapter 2, Section H.1).

The atmosphere acquires moisture by evaporation from oceans, lakes, rivers, and damp soil or from moisture transpired from plants. Taken together, these are often referred to as *evapotranspiration* and the mechanisms involved will now be discussed in detail.

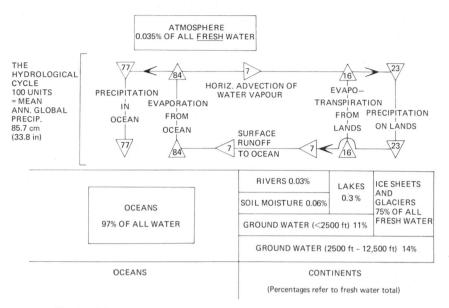

Fig. 2.1. The hydrological cycle and water storage of the globe. The exchanges in the cycle are referred to 100 units which equal the mean annual global precipitation of 85.7 cm (33.8 in.). The percentage storage figures for atmospheric and continental water are percentages of all fresh *water. The saline ocean waters make up 97% of* all *water (from More, 1967).*

A.   EVAPORATION

Evaporation occurs whenever energy is transported to an evaporating surface if the vapor pressure in the air is below the saturated value (e_s). As discussed in Chapter 1, Section B.2, the saturation vapor pressure increases with temperature. The change in state from liquid to vapor requires energy to be expended in overcoming the intermolecular attractions of the water particles. This energy is generally provided by the removal of heat from the immediate surroundings causing an apparent heat loss (*latent heat*) and a consequent drop in temperature. The latent heat of vaporization to evaporate 1 gram of water at 0°C is 600 cal and at 100°C it is 540 cal. Conversely, condensation releases this heat, and the temperature of an air mass in which condensation is occurring is increased as the water vapor reverts to the liquid state. The diurnal range of temperature is often moderated by damp air conditions, when evaporation takes place during the day and condensation at night.

Viewed another way, evaporation implies an addition of kinetic energy to individual water molecules and, as their velocity increases, so the chance of individual surface molecules escaping into the atmosphere becomes greater. As the faster molecules will generally be the first to escape, so the average energy (and therefore temperature) of those composing the remaining liquid will decrease and the quantities of energy required for their continued release become correspondingly greater. In this way evaporation decreases the temperature of the remaining liquid by an amount proportional to the latent heat of evaporation.

The rate of evaporation depends on a number of factors. The two most important are the difference between the saturation vapor pressure at the water surface and the vapor pressure of the air, and the existence of a continual supply of energy to the surface. Wind velocity can also affect the evaporation rate because the wind is generally associated with the importation of fresh, unsaturated air which will absorb the available moisture.

Water loss from plant surfaces, chiefly leaves, is a complex process termed *transpiration*. It occurs when the vapor pressure in the leaf cells is greater than the atmospheric vapor pressure, and is vital as a life function in that it causes a rise of plant nutrients from the soil and cools the leaves. The cells of the plant roots can exert an osmotic tension of up to about 15 atmospheres upon the water films between the adjacent soil particles. As these soil water films shrink, however, the tension within them increases. If the tension of the soil films exceeds the osmotic root tension the continuity of the plant's water supply is broken and wilting occurs. Transpiration is controlled by the atmospheric factors which determine evaporation as well as by plant factors such as the stage of plant growth, leaf area, and leaf temperature, and also by the amount of soil moisture (see Chapter 7, Section B.3). It occurs mainly during the day when the *stomata* (that is, small pores in the leaves) through which transpiration takes place are open. This opening is determined primarily by light intensity. Transpiration naturally varies greatly with season, and during the winter months in mid-latitudes conifers lose only 10–18% of their total annual transpiration losses and deciduous trees less than 4%.

In practice it is difficult to separate water evaporated from the soil, *intercepted moisture* remaining on vegetation surfaces after precipitation and subsequently evaporated, and transpiration. For this reason evaporation is sometimes applied as a general term for all these, or, more correctly, the composite term evapotranspiration may be used.

Evapotranspiration losses from natural surfaces cannot be measured directly. There are, however, various indirect methods of assessment, as well as theoretical formulas. One approximate means of indirect measurement is based on the moisture balance equation:

$$\text{Precipitation} = \text{Runoff} + \text{Evapotranspiration} + \begin{bmatrix} \text{Change in} \\ \text{soil moisture storage} \end{bmatrix}$$

Essentially the method is to measure the percolation through an enclosed block of soil with a vegetation cover (usually grass) and to record the rainfall upon it. The block, termed a *lysimeter*, is weighed regularly so that weight changes unaccounted for by rainfall or runoff can be ascribed to evapotranspiration losses, provided the grass is kept short! The technique allows the determination of daily evapotranspiration amounts.

If the soil block is regularly irrigated so that the vegetation cover is always yielding the maximum possible evapotranspiration the water loss is called the potential evapotranspiration (or PE).[1] Assuming a constant soil moisture storage, *potential evapotranspiration* is calculated as the difference between precipitation and percolation. A simple evapotranspirometer installation is shown in Fig. 2.2; the double tank installation ensures that representative readings are obtained. Potential evapotranspiration forms the basis for one system of climate classification developed by C. W. Thornthwaite (see Appendix 1).

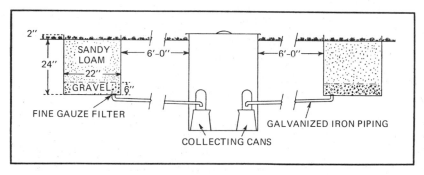

Fig. 2.2. An evapotranspirometer installation for calculating potential evapotranspiration losses. The double installation allows an average of the two results to be determined, giving a more reliable estimate (from Ward, 1963).

Theoretical methods for determining evaporation rates have followed two lines of approach. The first relates average monthly evaporation (E) from large water bodies to the mean wind speed (u) and the mean vapor pressure difference between the water surface and the air ($e_w - e_d$) in the form:

$$E = Ru(e_w - e_d),$$

where R is an empirical constant. This is termed the aerodynamic approach because it takes account of the factors responsible for removing vapor from the water surface. The second method is based on the energy budget. The *net balance* of solar and terrestrial radiation at the surface (R_N) is used for evaporation (E) and the conduction of heat to the atmosphere (H). A small proportion also heats the soil by day, but since nearly all of this is lost at night it can be disregarded. Thus,

$$R_N = LE + H,$$

[1] PE can be defined more generally as the water loss corresponding to the available energy.

where L is the latent heat of evaporation. Unfortunately, although the net radiation R_N can be measured with radiation instruments, H can only be estimated as a proportion of E. H/LE is referred to as Bowen's ratio (β) and is proportional to the ratio of the vertical gradients of temperature and vapor pressure between the surface and the level of instrumental measurements in the air. Evaporation is determined from an expression of the form:

$$E = \frac{R_N}{L(1 + \beta)}.$$

The most satisfactory method so far devised combines the two approaches. In this way H. L. Penman eliminated some of the unmeasurable quantities

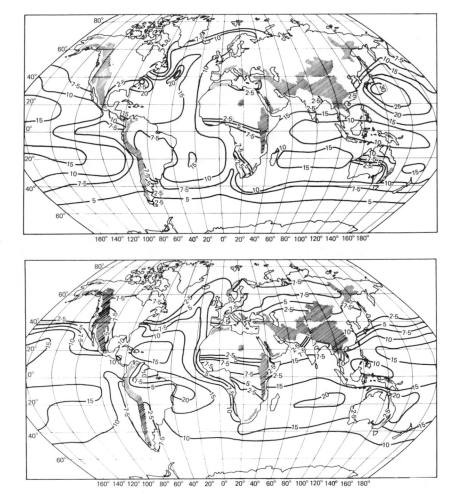

Fig. 2.3. Mean evaporation (cm) for January and July.

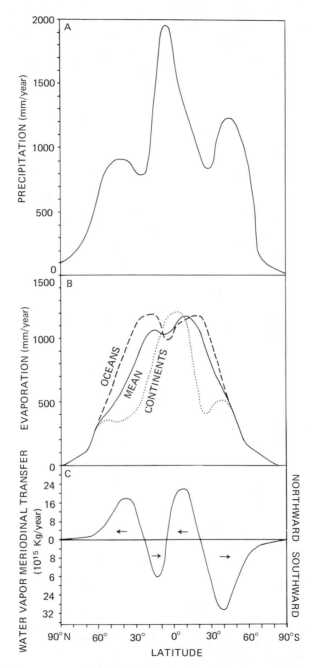

Fig. 2.4. The average annual latitudinal distribution of (A) precipitation (in mm); (B) evaporation (in mm); and (C) meridional transfer of water vapor (in 10¹⁵ kg) (mostly from Sellers, 1965).

and succeeded in expressing evaporation losses in terms of four meteorological elements which are regularly measured, at least in Europe and North
America. These are duration of sunshine (related to radiation amounts),
mean air temperature, mean air humidity, and mean wind speed (which
limit the losses of heat and vapor from the surface).

The relative roles of the factors which have been mentioned are illustrated
by the global pattern of evaporation (Fig. 2.3). Losses decrease sharply in
high latitudes where there is little available energy. In middle and lower
latitudes there are appreciable differences between land and sea (Fig. 2.4B).
Rates are naturally high over the oceans in view of the unlimited availability
of water, and the maximum rates occur in winter over the western Pacific
and Atlantic, where cold continental air blows across warm ocean currents.
On an annual basis maximum oceanic losses occur about 15°–20°N and
10°–20°S in the belts of the constant trade winds. The highest annual losses,
estimated to be about 200 cm (80 inches), are in the western Pacific and
central Indian Ocean near 15°S. (cf. Fig. 1.27; 100 Kcal/cm² is equivalent
to an evaporation of 170 cm of water/cm²). There is a subsidiary equatorial
minimum over the oceans mainly as a result of the lower wind speeds in the
doldrum belt and the proximity of the vapor pressure in the air to its saturation value, but the land maximum occurs more or less at the equator because
of the relatively high solar radiation receipts and the large transpiration
losses from the luxuriant vegetation of this region. The secondary maximum
over land in mid-latitudes is related to the strong prevailing westerly winds.
The other parts of Fig. 2.4, incorporated here for convenient comparison,
are discussed in later sections.

The annual evaporation over Britain, calculated by Penman's formula,
ranges from about 15 inches (38 cm) in Scotland to about 20 inches (50 cm)

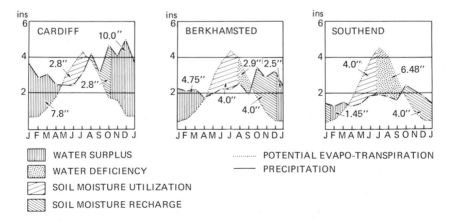

Fig. 2.5. The average annual moisture budget for stations in western, central and eastern
Britain determined by Thornthwaite's method (from Howe, 1956).

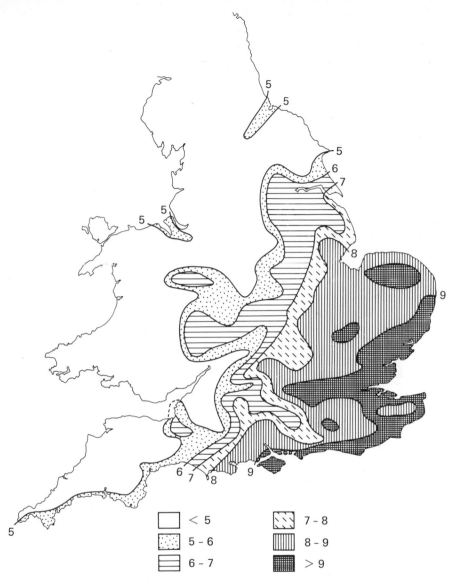

Fig. 2.6. *The average number of years in ten when irrigation is theoretically necessary for crops in England and Wales, based on Penman's formula (from Pearl et al., 1954) (Crown Copyright Reserved).*

in parts of south and southeast England. The annual potential evapotranspiration determined by Thornthwaite's method (based on mean temperature) is over 25 inches (64 cm) in most of southeastern England. Since this loss is concentrated in the period May–September there may be seasonal water deficits of 5–6 inches (12–15 cm) in these parts of the country (as shown in Fig. 2.5 for Southend), necessitating considerable use of irrigation water by farmers. Fig. 2.6 indicates that in southern and southeastern England it is necessary to irrigate in about nine years out of ten during the summer six months (April–September), assuming that the crop can extract $2\frac{1}{2}$ inches (6.4 cm) of moisture from the soil.

B. HUMIDITY

1. Moisture content. The moisture content of the atmosphere can be measured in a number of ways, apart from the vapor pressure, depending on which aspect the user wishes to emphasize. The total mass of water in a given volume of air, that is, the density of the water vapor, is one such measure. This is termed the *absolute humidity* (ρ_w) and is measured in grams per cubic meter (g/m³). Volumetric measurements are not greatly used in meteorology and more convenient is the *mass mixing ratio* (*W*). This is the mass of water vapor in grams per kilogram of dry air. For most practical purposes the specific humidity (*q*) is identical, being the mass of vapor per kilogram of air including its moisture.

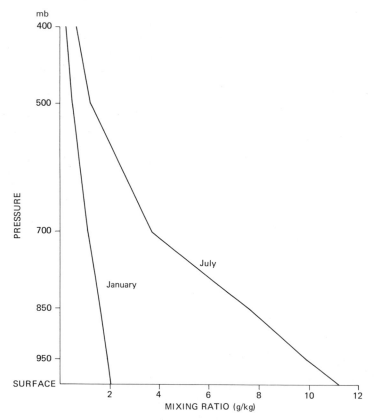

Fig. 2.7. The average vertical variation of atmospheric vapor content at Portland, Maine, between 1946 and 1955 (data from Reitan, 1960).

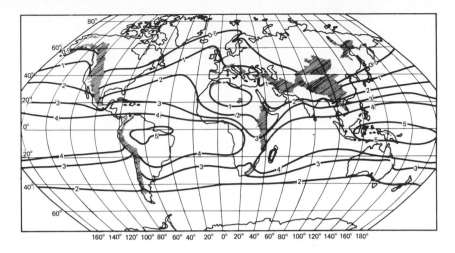

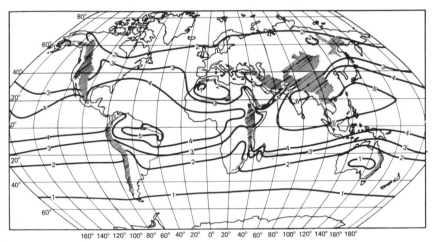

Fig. 2.8. Mean atmospheric water vapor content in January and July, 1951–55, in cm of precipitable water (after Bannon and Steele, 1960. Crown Copyright Reserved.)

The bulk of the atmosphere's moisture content is contained below 500 mb (5574 meters), as Fig. 2.7 clearly shows. It is also apparent that the seasonal effect is most marked in the lowest 3000 meters (10,000 ft) or so, that is, below about 700 mb. The global distribution of atmospheric vapor content in January and July is illustrated in Fig. 2.8. Over southern Asia during the summer monsoon an air column holds 5–6 cm of precipitable water, compared with less than 1 cm in tropical desert areas. Minimum values of 0.1–0.2 cm occur over high latitudes and continental interiors of the northern hemisphere in winter.

Another important measure is *relative humidity* (r), which expresses the actual moisture content of a sample of air as a percentage of that contained in the same volume of saturated air at the same temperature. The relative humidity is defined with reference to the mixing ratio, but it can be determined approximately in several ways:

$$r = \frac{W}{W_s} \times 100 \approx \frac{q}{q_s} \times 100 \approx \frac{e}{e_s} \times 100,$$

where the subscript s refers to the respective saturation values at the same temperature.

A further index of humidity is the dew-point temperature. This is the temperature at which saturation occurs if air is cooled at constant pressure without addition or removal of vapor. When the air temperature and dew point are equal the relative humidity is 100% and it is evident that relative humidity can also be determined from

$$\frac{e_s \text{ at dew point}}{e_s \text{ at air temperature}} \times 100.$$

The relative humidity of a parcel of air will obviously change if either its temperature or its mixing ratio is changed. In general the relative humidity varies inversely with temperature during the day, tending to be lower in the early afternoon and higher at night.

2. Moisture transport. It is sometimes overlooked that the atmosphere transports moisture horizontally as well as vertically. Fig. 2.4C illustrates the quantities which must be transported meridionally in order to maintain the required moisture balance at a given latitude (Precipitation minus Evaporation = Net horizontal transport of moisture into the air column). A prominent feature is the equatorward transport in low latitudes and the poleward transport in middle latitudes. The reader should inspect this diagram again in the light of the discussion of wind belts in Chapter 3, Section D.

At this point it is necessary to stress emphatically the fact that local evaporation is, in general, not the major source of local precipitation. For example, only 6% of the annual precipitation over Arizona and 10% of that over the Mississippi River basin is of local origin, the remainder being transported into these areas (that is, moisture advection). Even when moisture is available in the atmosphere over a specific region only a small portion of it is usually precipitated. This depends on the efficiency of the condensation and precipitation mechanisms, both microphysical and large-scale, which we shall now consider.

C. CONDENSATION

Condensation, the direct cause of all the various forms of precipitation, occurs under varying conditions which in one way or another are associated with a change in one of the linked parameters of air volume, temperature, pressure, or humidity. For instance, condensation takes place when the temperature of the air is reduced but its volume remains constant and the air is cooled to dew point, or whenever the volume of the air is increased without addition of heat (this cooling takes place because *adiabatic expansion* (see Chapter 2, Section E) causes energy to be consumed through work), or when a joint change of temperature and volume reduce the moisture-holding capacity of the air to below its existing moisture content. The key to the understanding of condensation clearly lies in the fine balance that exists between these independent variables. Whenever the balance between one or more of them is disturbed beyond a certain limit condensation may result.

The most common circumstances favorable to the production of condensation are those producing a drop in air temperature; namely contact cooling, mixing of air masses of different temperatures, and dynamic cooling of the atmosphere. Contact cooling is produced, for example, within warm, moist air passing over a cold land surface. On a clear winter's night strong radiation will cool the surface very quickly and this surface cooling will gradually extend to the moist lower air, reducing the temperature to a point where condensation occurs in the form of dew, fog, or frost, depending on the amount of moisture involved, the thickness of the cooling air layer, and the dew-point value. When the latter is below 0°C it is referred to as the hoar frost-point if the air is saturated with respect to ice.

The mixing of the differing layers within a single air mass or of two differing air masses can also produce condensation. Fig. 2.9 indicates both how the mixing of a vertical air layer may produce a relative humidity

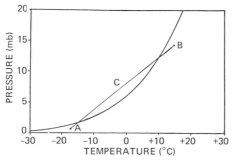
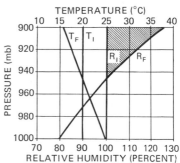

Fig. 2.9. The effect of air-mass mixing (from Petterssen, 1941). Left. The horizontal mixing of two unsaturated air masses A and B will produce one supersaturated air mass C. The saturation vapor pressure curve is shown. Right. The vertical mixing of an air mass having initially uniform temperature ($T_I = 20°C$) and humidity ($R_I = 100\%$) distributions, produces quite different final distributions with height (T_F and R_F) and a condensation level at about 947 mb.

exceeding 100%, and how the horizontal mixing of two air masses (*A* and *B*) having the given temperature and moisture characteristics may produce an air mass (*C*) which is over-saturated at the new temperature and will consequently become cloudy. Undoubtedly the most effective cause of condensation, however, is the dynamic process of adiabatic cooling. This involves an increase in volume of the mass of air due to a decrease of pressure, and the consequent lowering of its temperature towards dew point. It usually occurs when air is lifted to a lower pressure level without the addition of heat energy. In short, a volume increase involves work and the consumption of energy, thus reducing the heat available per unit volume and hence the temperature.

1. Condensation nuclei. It is very important to note that condensation occurs with the utmost difficulty in *clean* air; moisture must generally find a suitable surface upon which it can condense. If pure air is reduced in temperature below its dew point it becomes *supersaturated* (that is, relative humidity exceeding 100%). To maintain a pure water drop of radius 10^{-9} m (0.001 μ) requires a relative humidity of 320%, and for one of 10^{-7} m (0.1 μ) radius 101%.

Usually condensation occurs on a foreign surface which can be a land or plant surface, as is the case for dew or frost, while in the free air condensation begins around so-called *hygroscopic nuclei*. These particles can be dust, smoke, sulphur dioxide, salts (NaCl), or similar microscopic substances, the surfaces of which (like the weather enthusiast's seaweed!) have the property of *wettability*. Moreover, hygroscopic aerosols are soluble. This is very important since the saturation vapor pressure is less over a solution droplet (for example, sodium chloride or sulphuric acid) than over a pure water drop of the same size and temperature (Fig. 2.10). Indeed, condensation begins on hygroscopic particles before the air is saturated; in the case of sodium chloride nuclei at 78% relative humidity. The nuclei range in size from those with a radius of 0.001 μ, which are ineffective because of the high supersaturations required for condensation, to *giants* of over 10 μ which do not remain airborne for very long. Sea salts, which are particularly hygroscopic, enter the atmosphere principally by the bursting of air bubbles in foam, but fine soil particles and chemical combustion products raised by the wind are equally important sources of nuclei. On average, oceanic air contains one million condensation nuclei per liter (thousand cm³) and land air holds some five or six million.

Once they are initially formed, the process of growth of water droplets is far from simple and much remains to be explained. In the early stages small drops grow more quickly than large ones, but, as the size of a droplet increases its growth rate by condensation decreases, as shown in Fig. 2.11. The radial growth rate obviously slows down as the drop size increases because there is an increasingly greater surface area to add to with every

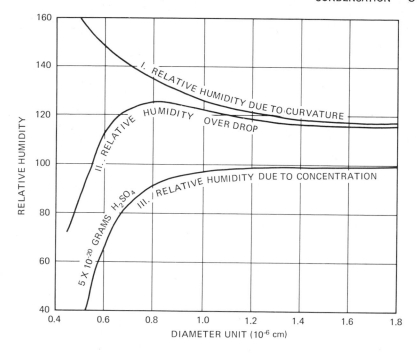

Fig. 2.10. The variation of relative humidity with droplet diameter. Curve I shows the effect due to curvature of the droplet and Curve III that due to a solution of 5 × 10⁻²⁰ g of sulphuric acid. Curve II is the net result of curvature and solution concentration (from Simpson, 1941).

increment of radius. However, the condensation rate is limited by the speed with which the released latent heat can be lost from the drop by conduction to the air and this heat reduces the vapor gradient. Moreover competition between droplets for the available moisture increasingly tends to reduce the degree of supersaturation.

Supersaturation in clouds very rarely exceeds 1% and, because the saturation vapor pressure is greater over a curved droplet surface than over a plane water surface, very small droplets (<0.1 μ radius) are readily evaporated (Fig. 2.10). In the early stages the nucleus size is important; for supersaturation of 0.05% a droplet of 1 μ radius with a salt nucleus of mass 10^{-13} grams reaches 10 μ in 30 minutes, whereas one with a salt nucleus of 10^{-14} grams would take 45 minutes. Later, when the dissolved salt has ceased to have significant effect, the radial growth rate becomes slow as a result of decreasing supersaturation (Fig. 2.10).

Fig. 2.11 illustrates not only the slow growth of droplets but also the immense size difference between cloud droplets (<1 μ to 50 μ radius) and raindrops (exceeding 1 mm diameter). These facts strongly suggest that the gradual process of condensation is inadequate to explain the rates of formation of raindrops which are often observed. For example, in most

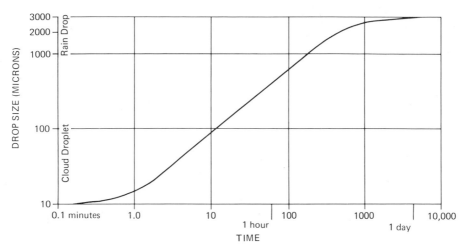

Fig. 2.11. Droplet growth by condensation (note the logarithmic scale).

clouds precipitation develops within an hour. It must be remembered too that falling raindrops undergo evaporation in the unsaturated air below the cloud base. A droplet of 0.1 mm radius evaporates after falling only 150 meters at a temperature of 5°C and 90% relative humidity, but a drop of 1 mm radius would fall 42 km before evaporating. It seems likely then that cloud droplets are not necessarily the immediate source of raindrops. This point is taken up again in Section D.

2. Clouds. The great variety of cloud forms necessitates a classification for purposes of weather reporting. The internationally adopted system

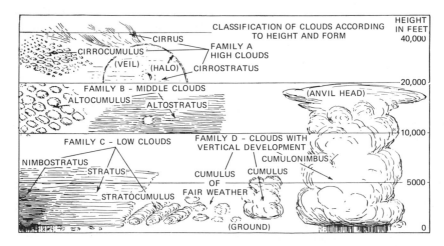

Fig. 2.12. The ten basic cloud groups classified according to height and form (from Strahler, 1965).

is based upon: (a) the general shape, structure, and vertical extent of the cloud, and (b) its altitude.

These primary characteristics are used to define the ten basic groups (or genera) as shown in Fig. 2.12. The high cirriform cloud is composed of ice crystals giving a generally fibrous appearance. Stratiform clouds are layer-shaped, while cumuliform ones have a heaped appearance and usually show progressive vertical development. Other prefixes are *alto-* for middle level (medium) clouds and *nimbo-* for low clouds of considerable thickness, which appear dark grey, and from which continuous rain is falling.

The height of the cloud base may show a considerable range for any of these types and varies with latitude. The approximate limits in thousands of meters for different latitudes are:

	Tropics	*Middle Latitudes*	*High Latitudes*
High cloud	Above 6	Above 5	Above 3
Medium cloud	2–7.5	2–7	2–4
Low cloud	Below 2	Below 2	Below 2

Following taxonomic practice, the classification subdivides the major groups into species and varieties with Latin names according to details of their appearance. The World Meteorological Organization has produced an *International Cloud Atlas* illustrating all of these types.

Other possible classifications of clouds take into account their mode of origin. For instance, a broad genetic grouping can be made according to the mechanism of vertical motion which produces condensation. Four categories are:

(a) gradual uplift of air over a wide area in association with a low-pressure system;
(b) thermal convection (on the local cumulus-scale);
(c) uplift by mechanical turbulence (*forced convection*);
(d) ascent over an orographic barrier (see Plate 3).

Group (a) includes a wide range of cloud types and will be discussed more fully in Chapter 4, Section E. In connection with thermal convection, which forms cumuliform clouds, it is worth noting that *towers* (castellanus) in cumulus or other clouds are due to thermals (rising air parcels) set up *within* the cloud as a result of the release of latent heat by condensation. Thermals gradually lose their impetus as mixing of cooler, drier air from the surroundings dilutes the more buoyant warm air. Group (c) includes fog, stratus, or stratocumulus and is important whenever air near the surface is cooled to dew point by conduction or nighttime radiation and the air is stirred by irregularities of the ground. The final group (d) could include stratiform or cumulus clouds produced by forced uplift of air over mountains. Hill fog is simply stratiform cloud enveloping high ground. A special and important category is the wave (lenticular) cloud which develops when air flows over hills setting up a wave motion in the air current downwind of the ridge (see Chapter 3, Section B.2). Clouds form in the crest of these waves if the air reaches its condensation level. (See Plates 4, 5, 6 and 7.)

A great deal of information on cloud amounts, especially in remote areas, and cloud patterns in relation to weather systems is now being provided by satellites of the United States' Tiros and Nimbus series. These investigations are supplying data that cannot be obtained by ground observations. For example, satellite pictures have shown a number of previously unrecognized patterns of cloud organization. Plate 8 illustrates one such feature, an actiniform or radiating system of cumulus clouds. Special classifications of these cloud patterns have been devised in order to file the immense number of photographs for future research investigations.

D. FORMATION OF PRECIPITATION

The puzzle of raindrop formation has already been briefly mentioned. The simple growth of cloud droplets is apparently an inadequate mechanism by itself, but if this is the case then more complex processes have to be envisaged.

Numerous early theories of raindrop growth have met with objections. For example, it was proposed that differently charged droplets could coalesce by electrical attraction, but it later appeared that distances between drops are too great and the difference between the electrical charges too small for it to happen. It was also suggested that large drops might grow at the expense of small ones, but observations showed that the distribution of droplet size in a cloud tends to maintain a regular pattern, with the average radius between about 10 μ to 15 μ and a few larger than 40 μ. Another proposal was based on the variation of saturation vapor pressure with temperature, such that if atmospheric turbulence brought warm and cold cloud droplets into close conjunction the supersaturation of the air with reference to the cold drop surfaces and the undersaturation with reference to the warm drop surfaces would cause the warm drops to evaporate and the cold ones to develop at their expense. However, except perhaps in some tropical clouds, the temperature of cloud droplets is too low for this differential mechanism to operate. Fig. 2.9 shows that below about 10°C the slope angle of the saturation vapor pressure curve is low. Another theory was that raindrops grow around exceptionally large condensation nuclei (such as have been observed in some tropical storms). Large nuclei, it is known, do experience a more rapid rate of initial condensation, but after this stage they are subject to the same limiting rates of growth that apply to all other atmospheric water drops.

The two main groups of current theories which attempt to explain the rapid growth of raindrops involve the growth of ice crystals at the expense of water drops, and the coalescence of small water droplets by the sweeping action of falling drops.

1. Bergeron–Findeisen theory. This theory forms an important part of the presently accepted mechanism of raindrop growth, and is based on the fact that the relative humidity of air is greater with respect to an ice surface than with respect to a water surface. As the air temperature falls below 0°C the atmospheric vapor pressure decreases more rapidly over an ice surface than over water (see Fig. 1.5). This results in the saturation vapor pressure over water becoming greater than that over ice, especially between temperatures of −5°C and −25°C where the difference exceeds 0.2 mb, and if ice crystals and supercooled water droplets exist together in a cloud the latter tend to evaporate and direct deposition takes place from the vapor on the ice crystals (this is often described by meteorologists as *sublimation*, which properly refers to direct evaporation from ice). Just as the presence of condensation nuclei is necessary for the formation of water droplets, so freezing nuclei are necessary before ice particles can form, usually at very low temperatures (about −15° to −25°C). Small water droplets can, in fact, be supercooled in pure air to −40°C before spontaneous freezing occurs. Freezing nuclei are far less numerous than condensation nuclei; for example there may be as few as 10 per liter at −30°C and probably rarely more than 1000. However, some become active at higher temperatures. Kaolinite, a common clay mineral, becomes initially active at −9°C and on subsequent occasions at −4°C. The origin of freezing nuclei has been the subject of much debate but it is generally considered that very fine soil particles are a major source. Another possibility is that meteoric dust provides the nuclei, although there seems to be no firm evidence of a relationship between meteorite showers and rainfall. Volcanic dust ejected into the upper stratosphere and troposphere during eruptions might be an additional terrestrial source.

Once minute ice crystals have formed they grow readily by deposition from vapor, with different hexagonal forms of crystal developing at different temperature ranges. The number of ice crystals also tends to increase progressively because small splinters become detached during growth by air currents and act as fresh nuclei. Ice crystals readily aggregate upon collision, due to their frequently branched (dendritic) shape, and tens of crystals may form a single snowflake. Temperatures between about 0°C and −5°C are particularly favorable to aggregation, because fine films of water on the crystal surfaces freeze when two crystals touch, binding them together. When the fall-speed of the growing ice mass exceeds the existing velocities of the air's upcurrents the snowflake falls, melting into a raindrop if it passes through a sufficient depth of air warmer than 0°C.

This theory seems to fit most of the observed facts, yet it is not completely satisfactory. Cumulus clouds over tropical oceans can give rain when they are only some 2000 meters (6500 ft) deep and the cloud-top temperature is 5°C or more. Even in middle latitudes in summer, precipitation may fall from cumuli which have no subfreezing layer (*warm clouds*). A suggested mechanism in such cases is that of droplet coalescence, which is discussed below.

Practical rainmaking has been based on the Bergeron theory with some success, which at least supports its principal points. The basis of such experiments is the freezing nucleus. Supercooled (water) clouds between −5°C and −15°C are *seeded* with especially effective nuclei, such as silver iodide or dry ice (CO_2), promoting the growth of ice crystals and encouraging precipitation. Some individually spectacular results have been achieved by such methods in Australia and the United States, but rainmaking cannot be claimed to be practicable on a routine basis at present. Its operation depends in any case on the presence of suitable supercooled clouds. When several cloud layers are present in the atmosphere, natural seeding may be important. For example, if ice crystals fall from high-level cirrostratus into altostratus composed of supercooled water droplets, the latter can grow rapidly by the Bergeron process and such situations may lead to extensive and prolonged precipitation.

2. Collision theories. Alternative raindrop theories use collision, coalescence, and sweeping as the drop growth generator. It was originally thought that atmospheric turbulence by making cloud particles collide would cause a significant proportion to coalesce. Unfortunately, it was found that particles might just as easily break up if subjected to collisions and it was also observed that there is often no precipitation from highly turbulent clouds. Langmuir offered a variation of this simple collision theory by pointing out that falling drops have terminal velocities directly related to their diameters such that the larger drops might overtake and absorb small droplets and that the latter might also be swept into the wake of the former, and be absorbed by them. Fig. 2.13 gives experimental results of the rate of growth of water drops by coalescence and also of ice particles by vapor deposition from an initial radius of 20 μ. Although coalescence is initially rather slow, the drop can reach 200 μ radius in just over 50 minutes. Calculations show that drops must exceed 19 μ in radius before they can coalesce with other droplets, smaller droplets being swept aside without colliding. The initial presence of a few larger drops calls for giant nuclei if the cloud top does not reach above the freezing level. However, if a few ice crystals are present at higher levels in the cloud (or if seeding occurs with ice crystals coming from higher cloud layers) they may eventually fall through the cloud as drops and the coalescence mechanism comes into action. Turbulence,

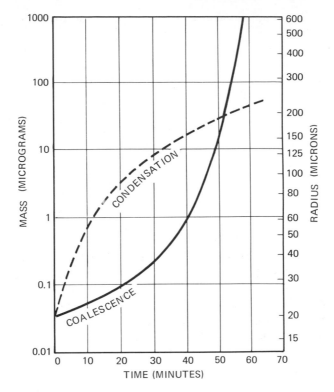

Fig. 2.13. Droplet growth by condensation and coalescence (from East and Marshall, 1954).

especially in cumuliform clouds, may serve to encourage collisions in the early stages. This coalescence process allows a more rapid growth than simple condensation can provide and is, in fact, fairly common in clouds in tropical maritime air masses, even in temperate latitudes.

3. Other types of precipitation. Rain has been discussed at length because it is the most common form of precipitation. Snow occurs when the freezing level is so near the surface that aggregations of ice crystals do not have time to melt before reaching the ground. Generally this means that the freezing level must be below 300 meters (1000 ft). Snow and rain are equally likely when the air temperature at the surface is about 1.5°C (34° to 35°F). Snow rarely occurs with an air temperature exceeding 4°C (39°F).

Soft hail pellets (roughly spherical, opaque grains of ice with much enclosed air) occur when the Bergeron process operates in a cloud with a small liquid water content and ice particles grow mainly by deposition of water vapor. Limited accretion of small supercooled droplets forms an aggregate of soft opaque ice particles 1 mm or so in radius. Showers of such pellets are quite common in winter and spring from cumulonimbus clouds.

Ice pellets may develop if the soft hail falls through a region of large liquid water content above the freezing level. Accretion forms a casing of clear ice around the pellet. Alternatively, an ice pellet consisting entirely of transparent ice may result from the freezing of a raindrop or the refreezing of a melted snowflake.

True hailstones are roughly concentric accretions of clear and opaque ice. The embryo is a raindrop carried aloft in an updraught and frozen. Successive accretions of opaque ice (rime) occur due to impact of super-cooled droplets which freeze instantaneously, whereas the clear ice (glaze) represents a wet surface layer, developed as a result of very rapid collection of supercooled drops in parts of the cloud with large liquid water content, which has subsequently frozen. A major difficulty in early theories was the necessity to postulate violently fluctuating upcurrents to give the observed banded hailstone structure, but a new thunderstorm model successfully accounts for this by demonstrating that the growing hailstones are cycled by the moving storm (see Chapter 4, Section D.2.*b*). On occasions hail-stones may reach giant size, weighing up to 1.5 lb each (recorded in 1928 at Potter's Bar, Nebraska). In view of their rapid fallspeeds hailstones may fall considerable distances with little melting.

E. ADIABATIC TEMPERATURE CHANGES

In Chapter 2, Section C it was noted that expansion of an air parcel involves work and the expenditure of energy. Thus, moving a parcel of air into a differ-ent pressure environment (without heat exchange with the surrounding air) produces a change of volume and of temperature (see Chapter 1, Section B also). Such a temperature change, involving no addition or subtraction of heat, is termed *adiabatic* and vertical displacements of air are obviously a major cause of such adiabatic temperature changes.

Near the earth's surface most processes of change are nonadiabatic (sometimes termed *diabatic*) because of the tendency of air to mix and modify its characteristics by lateral movement, turbulence, and related physical processes. When a parcel of air moves vertically the changes that take place often follow an adiabatic pattern because air is fundamentally a poor thermal conductor, and the air parcel as a whole tends to retain its own thermal identity which distinguishes it from the surrounding air masses. In some circumstances, on the other hand, mixing of air with its surroundings must be taken into account.

We may now consider the changes which occur when an air parcel rises and a decrease of pressure is accompanied by volume increase and tempera-ture decrease (see Chapter 1, Section B). The rate at which temperature decreases in a rising, expanding air parcel is called the adiabatic lapse rate.

If the upward movement of air does not produce condensation then the energy expended by expansion will cause the temperature of the mass to fall at what is called the *dry adiabatic lapse rate* (9.8°C/km or 5.4°F/1000 feet). However, prolonged reduction of the temperature invariably produces condensation, and when this happens latent heat is liberated, counteracting the dry adiabatic temperature decrease to a certain extent. It is therefore a distinguishing feature of rising and saturated (or precipitating) air that it cools at a slower rate (that is, the *saturated adiabatic lapse rate*) than air which is unsaturated. Another difference between the dry and saturated adiabatic rates is that whereas the former remains constant the latter varies with temperature. This is because air masses at higher temperatures are able to hold more moisture and on condensation, therefore, to release a greater quantity of latent heat. For high temperatures the saturated adiabatic lapse rate may be as low as 4°C/km (or 2.2°F/1000 feet), but this rate increases with decreasing temperatures, approaching 9°C/km (5°F/1000 feet) at −40°C (−40°F).

In all, three different lapse rates can be differentiated, two dynamic and one static. There is the environmental (or static) rate, which is the actual temperature decrease with height on any occasion, such as an observer ascending with a balloon would record. This is not an adiabatic rate, therefore, and may assume any form depending on local air temperature conditions. There are also the dynamic adiabatic dry and saturated lapse rates (or cooling rates) which apply to rising parcels of air moving through their environment. Close to the surface the vertical temperature gradient sometimes greatly exceeds the dry adiabatic lapse rate, that is, it is super-adiabatic. This is particularly common in arid areas in summer (see Table 1.3). Over most ordinary dry surfaces the lapse rate approaches the dry adiabatic value at an elevation of 100 meters or so.

The changing properties of moving air parcels can be conveniently expressed by plotting them as path curves on suitably constructed graphs. Such graphs are the tephigram (Fig. 2.14) and the adiabatic chart (Fig. 2.15). On these the properties of moving air can be compared with those of the surrounding air (plotted as environment curves) to enable inferences to be drawn regarding the stability of air parcels or layers (see Chapter 2, Section F).

The tephigram is composed of five properties:

(1) *Isotherms*, that is, lines of constant temperature (parallel lines from bottom left to top right).
(2) Dry adiabats (parallel lines from bottom right to top left).
(3) *Isobars*, that is, lines of constant pressure (slightly curved nearly horizontal lines).
(4) Saturated adiabats (curved lines sloping up from right to left).
(5) Saturation mixing ratio lines (those at a slight angle to the isotherms).

Fig. 2.14 shows, for example, that a saturated air parcel at 1000 mb pressure, with a temperature of 20°C, has a saturation mixing ratio of 15 g/kg. The same figure illustrates how the tephigram is used to determine the lifting condensation level—the level at which an air parcel becomes saturated if forced to rise. The point at which a line along a dry adiabat through the surface air temperature value intersects a line along a saturation mixing ratio line through the dew-point temperature is the condensation level. For an air temperature of 10°C and a dew point of 6°C at 1000 mb pressure, the condensation level is at 940 mb with a temperature of 5°C. The other lines on the chart are discussed in the next section.

Another type of thermodynamic diagram is the aerogram (Fig. 2.15). Here the isotherms are vertical instead of being skewed lines as on the tephigram, and the dry adiabats are slightly curved. The figure shows the parcel on the right (with a temperature of 25°C and 75% relative humidity at 1000 mb) rising and cooling dry-adiabatically to the condensation level (CL) and thereafter cooling at the saturated adiabatic rate.

Dry adiabats are also lines of equal *potential temperature* (*isentropes*), that is, the temperature which an air parcel would have if brought dry-adiabatically to the 1000-mb level. Potential temperature provides an impor-

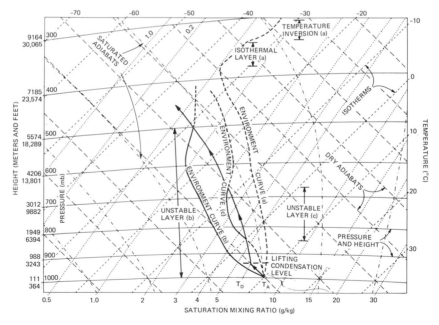

Fig. 2.14. A tephigram illustrating different cases of stable and unstable air. The arrowed line shows the path curve followed by a rising parcel of air with a surface temperature (T_A) of 10°C and a dew point (T_D) of 6°C at 1000 mb. The dashed portions of the individual environment curves are stable and the solid line sections are unstable. A description is given in the text.

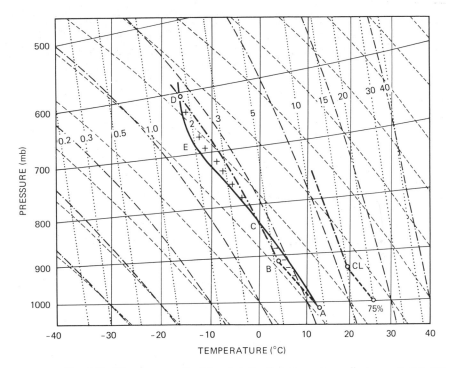

Fig. 2.15. The aerogram. On this adiabatic chart the main coordinates are temperature (°C) and pressure (mb). Saturation mixing ratio lines (g/kg) are dotted. Dry abiabats (dashed lines) and saturation adiabats (dashed and dotted) are also shown. A description is given in the text (from Petterssen, 1941).

tant yardstick for air-mass characteristics, since if the air is only affected by dry adiabatic processes the potential temperature remains constant. This helps to identify different air masses and indicates when latent heat has been released through saturation of the air mass or when nonadiabatic temperature changes have occurred.

F. AIR STABILITY AND INSTABILITY

The important characteristic of stable air is that if it is forced up or down it has a tendency to return to its former position once the motivating force ceases. Fig. 2.14 shows the reason for this, in that the environmental temperature curve (*a*) lies to the right of the path curve representing the lapse rate of an unsaturated air parcel cooling dry adiabatically when forced to rise. At any level the rising parcel is cooler and more dense than its surroundings and therefore tends to revert to its former level. Similarly, if the air is forced downwards it will gain in temperature at the dry adiabatic rate, always be warmer and less dense than the surrounding air, and tend to return to its former position (unless prevented from doing so). However, if the surrounding air has an environmental lapse rate (b) which exceeds that of the dry adiabat then the effect is reversed. When air is forced to rise it will cool at the dry adiabatic rate, will always be warmer than the air surrounding it, and tend to continue to rise. Similarly, if the air parcel is forced downwards it will always be colder than the adjacent air and there is no check on its downward progress until it reaches the surface. The characteristic of unstable air is a tendency to continue moving away from its original level when once initially set in motion. Environment curve (b) in Fig. 2.14 intersects the path curve at the 500-mb level. This is the level at which instability ceases and represents the general upper limit of cloud development. Neutral air, in contrast, stays at the level it has attained when the impelling movement ceases (that is, the path curves and the environment curves are identical).

A further possibility is illustrated by environment curve (c). The air is stable between 1000 mb and 850 mb. If the air is forced to rise, for example,

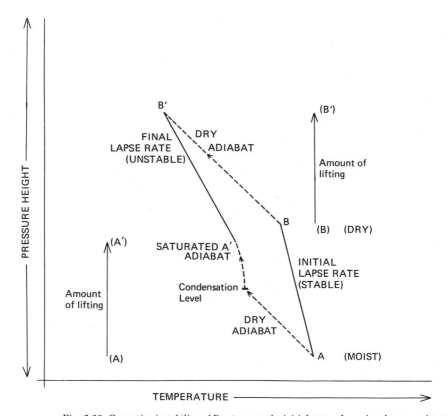

Fig. 2.16. Convective instability. AB represents the initial state of an air column; moist at A, dry at B. After uplift of the whole air column the temperature gradient A′B′ exceeds the saturated adiabatic lapse rate so that the air column is unstable.

by passage over a mountain range, to a height of 1500 meters or more (that is, above 850 mb) the path curve is now to the right of the environment curve and the air, being warmer than its surroundings, is free to rise up to 670 mb. This is termed conditional instability, as the development of instability is dependent on the relative humidity of the air. Since the environmental lapse rate is frequently between the dry and saturated adiabatic rates the state of conditional instability is a common one.

Fig. 2.15 illustrates a parcel of air rising and cooling dry-adiabatically from A to its condensation level at B, above which it cools at the saturated rate. The solid line $ACED$ shows the environmental temperature curve. Below C rising air is always cooler than its surroundings. If forced to rise above C the air parcel can then rise to almost 600 mb at D. Again this is a case of conditional instability.

These examples assume that a small air parcel is being displaced without any compensating air motion or mixing of the parcel with its surroundings. These assumptions may not hold good if there is active thermal convection, but the method is generally satisfactory for routine forecasting. A further consideration is that a deep layer of air may be displaced by vertical motion over an extensive topographic barrier. Fig. 2.16 shows a case where the air in the upper levels is less moist than that at lower levels. If the whole layer is forced bodily upwards the drier air at B follows the dry adiabatic rate, and so, for a while, will the air about A, but there eventually comes a time when the condensation level is reached, after which the lower layers of the rising air mass cool at the saturated adiabatic rate. This has the final effect of increasing the actual lapse rate of the total thickness of the raised layer, and, if this new rate is greater than that of the saturated adiabatic, the air layer becomes unstable and may overturn. This is termed convective (or potential) instability.

Subsidence usually results from either radiational cooling or an excess of horizontal convergence of air in the upper troposphere. Subsiding air generally moves with a vertical velocity of only 1–10 cm/sec (about 100–1000 ft/hr), although convectional conditions provide an exception (see Chapter 2, Section G). Subsidence can produce substantial changes in the atmosphere, and, for instance, if an air mass subsides about 300 meters (1000 ft) all average-size cloud droplets will usually be evaporated.

G. THUNDERSTORMS

In temperate latitudes probably the most spectacular example of moisture changes and associated energy releases in the atmosphere is the thunderstorm. Unusually great upward and downward movements of air are both the principal ingredients and motivating machinery of such storms. They occur: (*a*) as rising cells of excessively heated moist air; (*b*) along a *squall-line* in association with an air-mass discontinuity (see Chapter 4, Section

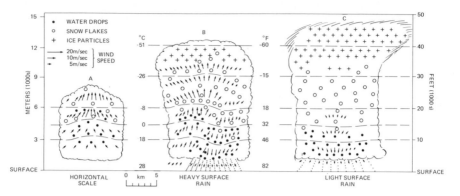

Fig. 2.17. The cycle of thunderstorm development. The arrows indicate the direction and speed of air currents. Left. The developing stage of the initial updraught. Center. The mature stage with updraughts and downdraughts. Right. The dissipating stage dominated by cool downdraughts. (after Byers and Braham. Adapted from Petterssen, 1958.)

D.2.b); or (c) in association with the triggering-off of conditional instability by uplift over mountains or by excessive local convergence (see Chapter 3, Section C.5).

The life cycle of a storm lasts only one to two hours, and begins when a parcel of air is either warmer than the air surrounding it or is actively under-cut by colder encroaching air. In both instances the air begins to rise and the embryo thunder cell forms as an unstable updraught of warm air (Fig. 2.17). As condensation begins to form cloud droplets, latent heat is released and the initial upward impetus of the air parcel is augmented by an expansion and a decrease in density until the whole mass becomes completely out of thermal equilibrium with the surrounding air. At this stage updraughts commonly reach 10 m/sec, and may exceed 30 m/sec. The constant release of latent heat continuously injects fresh supplies of heat energy which accelerate the updraught and do not permit it to slacken. The rise of the air mass will continue as long as its temperature remains greater (or, in other words, its density less) than that of the surrounding air.

Raindrops probably begin to develop rapidly when the ice stage (or freezing stage) is reached by the vertical build-up of the cell, but they do not immediately fall to the ground because the updraughts are able to support them. An early and now unacceptable theory of thunderstorm electricity postulated that when raindrops rupture (after reaching a size of about 5 mm diameter) the larger droplets retain a positive charge and the spray from the surface film of water consists of negative ions (Lenard effect). This spray was assumed to be carried upwards giving the cloud tops a negative charge while the drops fell and charged the lower parts positively. However, observations demonstrated the opposite distribution of charges in most thunderclouds. One possible explanation is that, since the earth is charged negatively and the ionosphere positively, charged droplets in the atmosphere have an induced positive charge on their lower side and negative charge on their upper side. As the droplets fall they brush aside ions which have the same sign as their lower surface and collect negative ions (ions, it will be recalled, are charged particles formed by radioactive gases from the earth or by the action of cosmic and ultraviolet rays). In this way the droplet becomes negatively charged and these falling droplets are naturally concentrated in the lower part of the cloud. This may explain part of the observed distribution of charges shown in Fig. 2.18.

Many current theories of the generation of electrical charges are based on the effects of freezing. A major contribution has been made along these lines by Latham and Mason, involving two related ideas. A supercooled droplet freezes inwards from its surface and this leads to a negatively charged warmer core (OH$^-$ ions) and a positively-charged colder surface due to the migration of H$^+$ ions outwards down the temperature gradient. When this soft hailstone ruptures during freezing small ice splinters carrying a

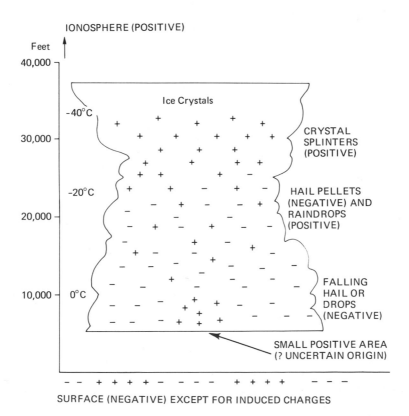

Fig. 2.18. The distribution of electrostatic charges in a thunder-cloud (after Mason, 1962).

positive charge are ejected by the ice shell and these are preferentially lifted to the top of the convection cell in updraughts. This theory helps to complete our understanding of the charge distribution in Fig. 2.18, which shows that the upper part of the cloud (above about the $-20°C$ isotherm) is positively charged. Equally, the negatively charged hail pellets fall towards the cloud base; this factor is probably more important than the collection of negative ions by drops mentioned previously. Another process by which the thundercloud charges may be created involves the collision between cold ice crystals and warmer pellets of soft hail. Previous accretion of supercooled droplets on the hail pellets produces an irregular surface which is warmed as the droplets release latent heat on freezing. The impacts of cold ice crystals on this irregular surface then generate negative charge while the colder crystals acquire positive charge. Again the effects of gravitational separation would lead to the observed distribution of charges. The origin of small positive areas near the cloud base (Fig. 2.18) is still under discussion.

Lightning commonly begins more or less simultaneously with precipitation downpours. It may occur between the lower part of the cloud and the ground (which locally has an induced positive charge). The first relatively weak (leader) stage of the flash bringing down negative charge from the cloud is met near the ground by a return stroke which rapidly takes positive charge upward along the already formed channel of ionized air. Just as the leader is neutralized by the return stroke, so the latter is neutralized in turn within the clouds. Subsequent leaders and return strokes drain higher regions of the cloud until its supply of negative charge is temporarily exhausted. Other more frequent flashes occur within a cloud or between clouds. The extreme heating and explosive expansion of air immediately round the path of the lightning sets up intense sound waves causing thunder to be heard.

Lightning is only one aspect of the atmospheric electricity cycle. During fine weather the earth's surface is negatively charged, the ionosphere positively charged. Atmospheric ions can conduct electricity down to the earth and hence a return supply must be forthcoming to maintain the observed electrical field. One source is the slow *point discharge*, from objects such as buildings and trees, of ions carrying positive charge induced by the negative thundercloud base. Similar upward currents occur above thunderstorm clouds. The other source, estimated to be smaller in its effect over the earth as a whole, is the upward transfer of positive charge by the much more rapid lightning strokes, leaving the earth negatively charged. The joint operation of these supply currents, in approximately 1800 thunderstorms over the globe at any instant, is thought to be sufficient to balance the air–earth leakage, and this number seems to agree reasonably well with observational data.

The middle or mature stage of a storm (Fig. 2.17B) is reached when the action of the falling precipitation causes frictional downdraughts of cold air. The cooling may be strengthened by evaporation from the falling drops and as the downdraughts gather momentum the cold air tends to spread out below the thunder cell in a wedge. Gradually, as the moisture of the cell is expended, the supply of released latent heat energy diminishes, the downdraughts progressively gain in power over the warm updraughts, and the cell dissipates.

To simplify the explanation, a thunderstorm with only one cell was illustrated. Usually storms are far more complex in structure and consist of several cells arranged in clusters of 2–8 km across and extending vertically up to over 8000 meters. Radar studies (see, for example, Plate 9) show that convective storms tend to move some 20° to the right of the mid-tropospheric wind direction through the growth of new cells on the right flank as old ones decay on the left flank. New cells are commonly initiated by the collision of the cold downdraughts from two mature cells some miles apart, triggering the ascent of displaced warm air. It is during such growth of new cumulo-nimbus cells that a tornado may develop. These narrow, intense whirls of air grow downwards from the cloud base. Tornadoes are a regular occurrence in spring in the midwestern United States, where they are generally associated with thunderstorms on squall-lines (Chapter 4, Section D.5).

H. PRECIPITATION CHARACTERISTICS AND TYPES

Strictly, *precipitation* refers to all liquid and frozen forms of water—rain, sleet, snow, hail, dew, hoar-frost, fog-drip, and rime (ice accretion on objects through the freezing on impact of supercooled fog droplets)—but, in general, only rain and snow make significant contributions to precipitation totals. In many parts of the world the term rainfall can be used interchangeably with precipitation. The data in the following section refer to rainfall, since snowfall is less easily measured with the same degree of accuracy.

1. Precipitation characteristics. There are many measures by which the various attributes of precipitation can be described, both in the long term and from the point of view of individual storms. Traditionally such long-term measures as mean annual precipitation, annual variability, and year-to-year trends have been of great interest to the geographer, and these statistical measures are treated in the concluding chapter (see Chapter 8, Section A). However, particularly in terms of hydrological considerations, the characteristics and relationships of individual rainstorms are being studied increasingly, and it is possible here to point to some of their commonly observed features. Weather observations usually indicate the amount, duration, and frequency of precipitation and these enable other derived characteristics to be determined. Three of these are discussed on the following pages.

(a) Rainfall intensity. The intensity (=amount/duration) of rainfall during an individual storm, or a still shorter period, is of vital interest to
hydrologists and water engineers concerned with flood forecasting and
prevention, as well as to conservationists dealing with soil erosion. Chart
records of the rate of rainfall (*hyetograms*) are necessary to assess intensity,
which varies markedly with the time interval selected. Average intensities
for short periods (thunderstorm-type downpours) are much greater than
those for longer time intervals as Fig. 2.19 illustrates for Washington, D.C.

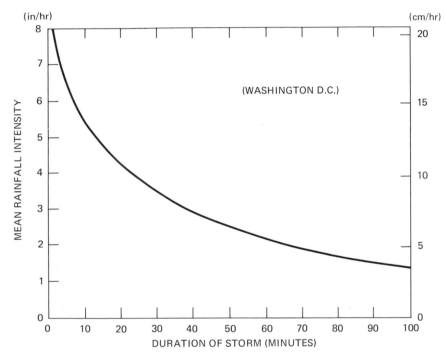

*Fig. 2.19. Generalized relationship between precipitation intensity and duration for Wash-
ington, D.C. (after Yarnell, 1935).*

In the case of extreme rates at different points over the earth (Fig. 2.20)
the record intensity over 10 minutes is approximately three times that for
100 minutes, and the latter exceeds by as much again the record intensity
over 1000 minutes (that is, 16½ hours). High-intensity rain is associated
with increased drop size rather than an increased number of drops. For
example, with precipitation intensities of 0.1, 1.3, and 10.2 cm/hr (that is,
0.05, 0.5, and 4.0 in./hr), the most frequent raindrop diameters are 0.1, 0.2,
and 0.3 cm, respectively. The occurrence of daily amounts exceeding 1.3 cm
(0.5 in.) is considered to be important for gully erosion in North America.
Such falls account for 90% of the annual rainfall on the Gulf coast compared
with only 20% in the Great Basin.

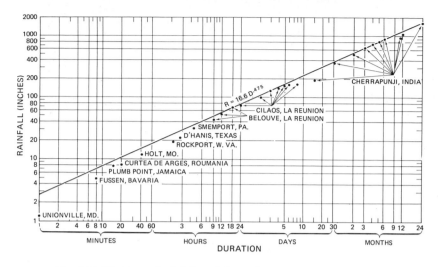

Fig. 2.20. World record rainfalls and the envelope of expected extremes at any place. The equation of the envelope line is given, together with the state or country where each record was established (from Paulhus, 1965).

(b) Areal extent of a rainstorm. The rainfall totals received in a given time interval vary according to the size of the area which is considered, showing a relationship analogous to that of rainfall duration and intensity. The maximum 24-hour rainfalls over areas of different extent in the United States (up to 1960) were as follows:

Sq. miles	Cm	In.
10 (25.9 km²)	98.3	38.7
10^2	89.4	35.2
10^3	76.7	30.2
10^4	30.7	12.1
10^5	10.9	4.3

(*After Gilman, 1964.*)

Fig. 2.21, based on data of this type, illustrates the maximum rainfall to be expected for a given storm area and given duration in the United States.

(c) Frequency of rainstorms. Another useful item is the average time-period within which a rainfall of specified amount or intensity can be expected to occur once. This is known as the *recurrence interval* or *return period*. Fig. 2.22 gives this type of information on rainfall amount and duration for Cleveland, Ohio. From this it would appear that a 10-minute

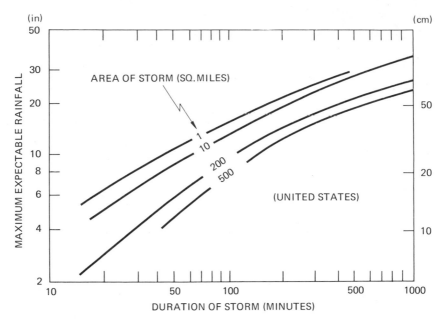

Fig. 2.21. Enveloping depth/duration curves of maximum rainfall for areas of under 500 square miles in the United States (after Berry, Bollay, and Beers, 1945).

fall of 2.5 cm (1 in.) is likely about once every twenty years, a 24-hour fall of 10 cm (4 in.) about once every thirty years. However, this *average* return period does not mean that such falls necessarily occur in the twentieth and thirtieth years, respectively, of a selected period. Indeed, they might occur in the first! These estimates require long periods of observational data, but the approximately linear relationships shown by such graphs are of great practical significance for the design of flood-control systems.

2. Precipitation types. The above material can now be related to that of the previous sections in a discussion of precipitation types. A convenient starting point is the usual division into three main types—convective, cyclonic, and orographic precipitation—according to the assumed mode of uplift of the air. Essential to this analysis is some knowledge of storm systems. These are treated in later chapters and the newcomer to the subject may prefer to read the following in conjunction with them.

(a) Convective type precipitation. This is associated with towering cumulus (cumulus congestus) and cumulonimbus clouds. Three subcategories can be distinguished according to their degree of spatial organization.

(*i*) Scattered convective cells develop through strong heating of the land surface in summer, especially when low upper tropospheric tempera-

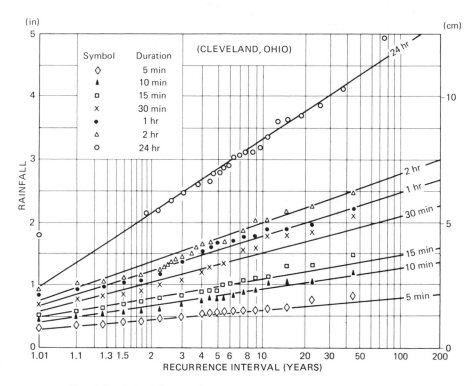

Fig. 2.22. Rainfall/duration/frequency curves for Cleveland, Ohio, 1902–1947 (after Linsley and Franzini, 1955).

tures facilitate the release of conditional or convective instability (see Chapter 2, Section F). Precipitation, often including hail, is of the thunderstorm type, although thunder and lightning do not necessarily occur. Limited areas, of the order of 20 to 50 km², are affected by the individual heavy downpours, which generally last for about $\frac{1}{2}$ to 1 hr.

(*ii*) Showers of rain, snow, or soft hail pellets may form in cold, moist unstable air passing over a warmer surface. Convective cells moving with the wind can produce a streaky distribution of precipitation parallel to the wind direction, although over a period of several days the variable paths and intensities of the showers tend to obscure this pattern. The cells may also be organized into a belt, some hundreds of kilometers long and 40–100 km (25–60 miles) wide, perpendicular to the airflow in association with cold fronts and squall lines (see Chapter 4, Section D.2.*b*). Hence the precipitation is widespread, though of brief duration at any locality.

(*iii*) In tropical cyclones cumulonimbus cells become organized about the vortex in spiralling bands (see Chapter 6, Section B.2.*a*). Particularly in the decaying stages of such cyclones, typically over land, the rainfall can be very heavy and prolonged, affecting areas of thousands of square kilometers.

(b) Cyclonic type precipitation. Precipitation characteristics vary according to the type of low-pressure system and its stage of development, but the essential mechanism is ascent of air through horizontal convergence of airstreams in an area of low pressure (see Chapter 3, Section C.5). In extratropical depressions this is reinforced by uplift of warm, less-dense air along an air-mass boundary (see Chapter 4, Section D.2). Such depressions give moderate and generally continuous precipitation over very extensive areas as they move usually eastward in the westerly wind belts between about 40° and 65° latitude. The precipitation belt in the forward section of the storm is associated with multilayered cloud and can give continuous light to moderate precipitation at localities in its path for 6 to 12 hr, whereas the belt in the rear gives a shorter period of showery precipitation often accompanied by thunder. These sectors are, therefore, sometimes distinguished in precipitation classifications, and a more detailed breakdown is illustrated in Table 5.4. Polar lows (see Chapter 4, Section G.3) combine the effects of airstream convergence and convective activity of category $a(ii)$, above, whereas troughs in the equatorial low-pressure area give convective precipitation as a result of airstream convergence in the easterlies (see Chapter 6, Section B.1).

(c) Orographic precipitation. Orographic precipitation is commonly regarded as a distinct type, but this requires careful qualification. Mountains are not especially efficient in causing moisture to be removed from airstreams crossing them, yet because precipitation falls repeatedly in more or less the same locations the cumulative totals are large. Orography, dependent on the alignment and size of the barrier, may (i) trigger conditional or convective instability by giving an initial upward motion or by differential heating of the mountain slopes, (ii) increase cyclonic precipitation by retarding the rate of movement of the depression system, (iii) cause convergence and uplift through the funnelling effects of valleys on airstreams. In mid-latitude areas where precipitation is predominantly of cyclonic origin, orographic effects tend to increase both frequency and intensity of winter precipitation, whereas during summer and in continental climates with a higher condensation level the main effect of relief is the occasional triggering of intense thunderstorm-type precipitation. It is perhaps helpful to distinguish between orographic precipitation, in the strict sense (that is, that which occurs over high ground when none is falling on the surrounding plains), and the *orographic component* of the total precipitation which results from the effect of orography on the primary convective and cyclonic mechanisms.

Two special cases of orographic effects may be mentioned. One is the general influence of surface friction which causes stratus or stratocumulus to form through turbulent uplift. Only light precipitation (drizzle, light rain, or snow grains) is to be expected under these circumstances. The other case arises through frictional slowing down of an airstream moving inland over the coast. A particular instance of the convergence and uplift which this may initiate has been reported by Bergeron. During a 24-hour period in October 1945, a west-southwesterly airstream over Holland produced a belt of precipitation (3 cm or more)—a result of frictional convergence and uplift—in crossing the narrow zone of coastal sand dunes only a few meters high. Over the remainder of that virtually flat country a series of lee waves developed in the tropospheric airflow downwind from the coast (see Fig. 3.7, for example), and these gave a series of transverse (north-south) bands of precipitation up to 2 cm in amount. On the following day the surface flow had changed little, but a temperature decrease from $-20°C$ to $-28°C$ at the 500-mb level altered the vertical stability so that the lee waves broke down and the precipitation distribution showed convective streaks, up to 4 cm per day, parallel to the surface wind direction.

3. Regional variations in the altitudinal maximum of precipitation. The increase of precipitation with height is a widespread characteristic. In middle and higher latitudes this is particularly pronounced on west-coast mountain ranges, but is less apparent inland. In North America the Coast Ranges, on the one hand, compared with the Rockies on the other, clearly demonstrate this effect (see Chapter 5, Section B.2). In western Britain with mountains of about 1000 meters, the maximum falls are recorded to leeward of the summits. This probably reflects the general tendency of air to go on rising for a while after it has crossed the crestline and the time-lag involved in the precipitation process after condensation. Over narrow uplands the horizontal distance may allow insufficient time for maximum cloud build-up and the occurrence of precipitation. However, a further factor may be the effect of eddies, set up in the airflow by the mountains, on the catch of rain gauges. Studies in Bavaria at the Hohenpeissenberg observatory show that standard rain gauges may overestimate amounts by about 10% on lee slopes and underestimate them by 14% on windward slopes.

In the tropics and subtropics the maximum precipitation occurs below the higher mountain summits, from which level it decreases upwards towards the crests. Observations are generally sparse in the tropics, but numerous records from Java show that the average elevation of greatest precipitation is approximately 1200 meters (4000 ft). Above about 2000 meters the decrease in amounts becomes quite marked. Similar features are reported from Hawaii and, at a rather higher elevation, on mountains in East Africa (see Chapter 6, Section D.3). Fig. 2.23A shows that, despite the wide range of records for individual stations, this effect is clearly apparent along the Pacific flank of the Guatemalan Highlands. Further north along the coast, the occurrence of a precipitation maximum below the mountain crest is observed in the Sierra Nevadas, despite some complication introduced by the shielding effect of the Coast Ranges (Fig. 2.23B), but in the Olympic Mountains of Washington precipitation increases right up to the summits (Fig. 2.23C). As has been previously mentioned, precipitation catches on mountain crests may underestimate the actual precipitation due to the effect of eddies, and this is particularly true where much of the precipitation falls in the form of snow which is very susceptible to drifting.

One explanation of this orographic difference between tropical and temperate rainfall is based on the concentration of moisture in a fairly shallow layer of air near the surface in the tropics (see Chapter 6, Section A). Much of the orographic precipitation seems to derive from warm clouds (particu-

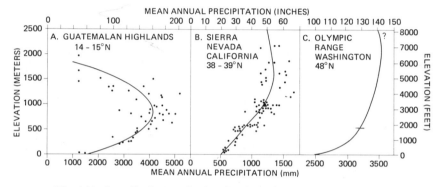

Fig. 2.23. *Generalized curves showing the relationship between elevation and mean annual precipitation for west-facing mountain slopes in Central and North America. The dots give some indication of the wide scatter of individual precipitation readings (adapted from Hastenrath, 1967 and Armstrong and Stidd, 1967).*

larly cumulus congestus), composed of water droplets, which commonly have an upper limit at about 3000 meters. It is probable that the height of the maximum precipitation zone is close to the mean cloud base since the maximum size and number of falling drops will occur at that level. Thus, stations located above the level of mean cloud base will receive only a proportion of the orographic increment. In temperate latitudes much of the precipitation, especially in winter, falls from stratiform cloud which commonly extends through a considerable depth of the troposphere. In this case there tends to be a smaller fraction of the total cloud depth below the station level. These differences according to cloud type and depth are apparent even on a day-to-day basis in middle latitudes, as has been shown by detailed studies in the Bavarian Alps. Seasonal variations in the altitude of the mean condensation level and zone of maximum precipitation are similarly observed. In the mountains of central Asia (the Pamirs and Tien Shan), for instance, the maximum is reported to occur at about 1500 meters (5000 ft) in winter and at 3000 meters (9900 ft) in summer.

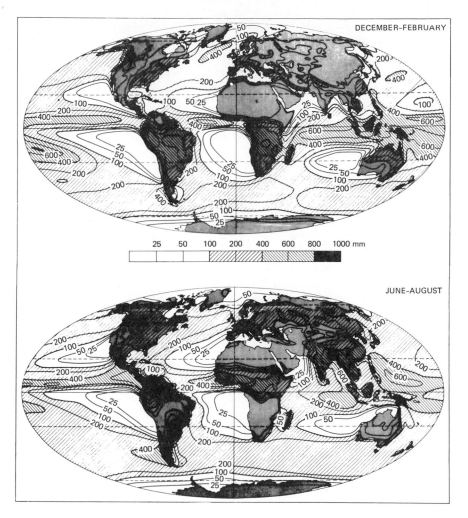

Fig. 2.24. Mean global precipitation (in mm) for the periods December–February and June–August (from Möller, 1951).

4. The world pattern of precipitation. A glance at the maps of precipitation amount for December–February and June–August (Fig. 2.24) indicates that the distributions are considerably more complex than those, for example, of mean temperature (see Fig. 1.14). Comparison of Fig. 2.24 with the meridional profile of average precipitation for each latitude (see Fig. 2.4A) brings out the marked longitudinal variations that are superimposed on the zonal pattern. The latter has three main features: an equatorial maximum which, like the thermal equator, is slightly displaced into the northern hemisphere; very low totals in high latitudes; and secondary minima in subtropical latitudes. Fig. 2.24 demonstrates why the subtropics do not appear as particularly dry on the meridional profile in spite of the known aridity of the subtropical high pressure areas (see Chapter 3, Section C.3 and Chapter 5, Section D.2). In these latitudes the eastern sides of the continents receive considerable rainfall in summer.

In view of the complex controls involved, no brief explanation of these precipitation distributions can be satisfactory. Various aspects of selected precipitation regimes are examined in Chapters 5 and 6, after consideration of the fundamental ideas about atmospheric motion, air masses and frontal zones. A classification of wind belts and precipitation characteristics is outlined in Appendix 1, Section C. It must suffice at this stage simply to note the factors which have to be taken into account in studying Fig. 2.24:

   (i) The limit imposed on the maximum moisture content of the atmosphere by air temperature. This is important in high latitudes and in winter in continental interiors.

  (ii) The major latitudinal zones of moisture influx due to atmospheric advection. This in itself is a reflection of the global wind systems and their disturbances (that is, the converging trade wind systems and the cyclonic westerlies, in particular).

 (iii) The distribution of the land masses. It is noteworthy that the southern hemisphere lacks the vast, arid, mid-latitude continental interiors of the northern. The oceanic expanses of the southern hemisphere allow the mid-latitude storms to increase the zonal precipitation average for 45°S by about one third compared with that of the northern hemisphere for 50°N (see Fig. 2.4A). Another major non-zonal feature is the occurrence of the monsoon regimes, especially in Asia.

 (iv) The distribution of mountain areas with respect to the locally prevailing winds.

3

Atmospheric Motion

The atmosphere acts somewhat in the same manner as a gigantic heat engine in which the constantly maintained difference in temperature existing between the poles and the equator provides the energy supply necessary to drive the planetary atmospheric circulation. The conversion of the heat energy into kinetic energy to produce motion must involve rising and descending air, but vertical movements are generally much less in evidence than horizontal ones, which may cover vast areas and persist for periods of a few days to several months.

Before considering these global aspects, however, it is important to look at the immediate controls on air motion. The pressure difference between the earth's surface and higher levels in the atmosphere might be expected to cause the atmosphere to escape, yet this does not happen because of the earth's gravitational field. The upward decrease in air pressure is balanced by the downward force of gravity producing a state referred to as hydrostatic equilibrium. This balance, together with the general stability of the atmosphere and its shallow depth, greatly limits vertical air motion. Average horizontal wind speeds are of the order of one hundred times greater than average vertical movements, though individual exceptions occur—particularly in convective storms.

A. LAWS OF HORIZONTAL MOVEMENT

There are four controls on the horizontal movement of air near the earth's surface: pressure gradient force, Coriolis force, centripetal acceleration and frictional forces. Of these, the Coriolis force and centripetal acceleration strictly are fictitious, but it is convenient to ascribe certain effects on air movement over the rotating earth's surface to such forces.

1. The pressure gradient force. The pressure gradient force has components in the vertical and horizontal planes but, as already noted, the vertical component is more or less in balance with the force of gravity. Spatial differences in pressure can be due to thermal or mechanical causes (often not easily distinguishable), and these differences control the horizontal movement of an air mass. In effect the pressure gradient serves as the motivating force which causes the movement of air away from areas of high pressure and towards areas where it is lower, although other forces prevent air from moving directly across the isobars (lines of equal pressure). The pressure gradient force per unit mass is expressed mathematically as

$$-\frac{1}{\rho}\frac{dp}{dn},$$

where ρ = air density and dp/dn = the horizontal gradient of pressure. Hence the closer the isobar spacing the more intense is the pressure gradient and the greater the wind speed. The pressure gradient force is also inversely proportional to air density and this relationship is of particular importance in understanding the behavior of upper winds.

2. The earth's rotational deflective (Coriolis) force. The Coriolis force arises from the fact that the movement of masses over the earth's surface is usually referred to a moving coordinate system (that is, the latitude and longitude grid which rotates with the earth). The simplest way to begin to visualize the manner in which this deflecting force operates is to picture a rotating disc on which moving objects are deflected. Fig. 3.1 shows the effect of such a deflective force operating on a mass moving outward from the center of a spinning disc. The body follows a straight path in relation to a fixed frame of reference (for instance, a box which contains a spinning disc), but viewed relative to coordinates rotating with the disc the body swings to the right of its initial line of motion. This effect is readily demonstrated if a pencil line is drawn across a white disc on a rotating turntable. Fig. 3.2 illustrates a case where the movement is not from the center of the turntable and the object possesses an initial momentum in relation to its distance from the axis of rotation. In the analogous case of the rotating earth (with rotating reference coordinates of latitude and longitude), there is apparent deflection of moving objects to the right of their line of motion in the northern hemisphere and to the left in the southern hemisphere, as viewed by observers on the earth. The deflective force (per unit mass) is expressed by:

$$-2\omega V \sin \phi,$$

where ω = the angular velocity of spin ($15°$/hr or $2\pi/24$ radians/hr for the earth = 7.29×10^{-5} radians/sec); ϕ = the latitude and V = the velocity of the mass. $2\omega \sin \phi$ is referred to as the Coriolis parameter (f).

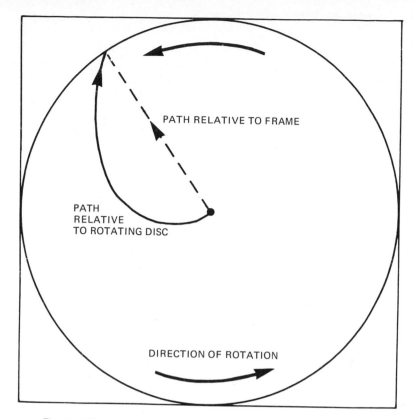

PATH RELATIVE TO FRAME

PATH
RELATIVE
TO ROTATING DISC

DIRECTION OF ROTATION

Fig. 3.1. The Coriolis deflecting force operating on a body moving outward from the center of a rotating turntable.

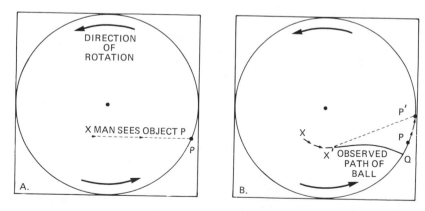

DIRECTION
OF
ROTATION

X MAN SEES OBJECT P

P

A.

P'

X

P

X' OBSERVED
PATH OF
BALL

Q

B.

Fig. 3.2. The Coriolis deflecting force on a rotating turntable. A. A man at X sees the object P and attempts to throw a ball towards it. Both locations are rotating anticlockwise. B. The man's position is now X' and the object is at P'. To the man, the ball appears to follow a curved path and lands at Q. The man overlooked the fact that P was moving to his left and that the path of the ball would be affected by the initial impetus due to the man's own rotation.

The magnitude of the deflection is directly proportional to: (*a*) the horizontal velocity of the air (that is, air moving at 11 m/sec (25 mph) having half the deflective force operating on it as that moving at 22 m/sec (50 mph)); and (*b*) the sine of the latitude (sin 0° = 0; sin 90° = 1). The effect is thus a maximum at the poles (that is, where the plane of the deflecting force is parallel with the earth's surface) and decreases with the sine of the latitude becoming zero at the equator (that is, where there is no component of the deflection in a plane parallel to the surface). Values of *f* vary with latitude as follows:

Latitude	0°	10°	20°	43°	90°
f (10^4/sec)	0	0.25	0.50	1.00	1.458

The Coriolis force always acts at right angles to the direction of the air motion to the right in the northern hemisphere and at right angles to the direction of the air motion to the left in the southern.

3. The geostrophic wind. Observations in the *free atmosphere* (above the level affected by surface friction at about 500 to 1000 meters) show that the wind blows more or less at right angles to the pressure gradient (that is, parallel to the isobars) with, for the northern hemisphere, the high-pressure core on the right and the low-pressure on the left when viewed downwind. This implies that for steady motion the pressure gradient force is exactly balanced by the Coriolis deflection acting in the diametrically opposite direction (Fig. 3.3). The wind in this idealized case is called a *geostrophic wind*, the velocity (V_g) of which is given by the following formula:

$$V_g = \frac{1}{2\omega \sin \phi \, \rho} \cdot \frac{dp}{dn},$$

where dp/dn = the pressure gradient. The velocity is thus inversely dependent on latitude, such that the same pressure gradient which will be associated with geostrophic wind speeds of 15 m/sec (34 mph) at latitude 43° will produce a velocity of only 10 m/sec (23 mph) at latitude 90°. Except in low latitudes, where the Coriolis deflection approaches zero, the geostrophic wind is a close approximation to the observed air motion in the free atmosphere. Since pressure systems are rarely stationary this fact implies that air motion must continually change towards a new balance. In other words mutual adjustments of the wind and pressure fields are constantly taking place. The common cause-and-effect argument that a pressure gradient is formed and air begins to move towards low pressure before coming into geostrophic balance is an unfortunate oversimplification of reality.

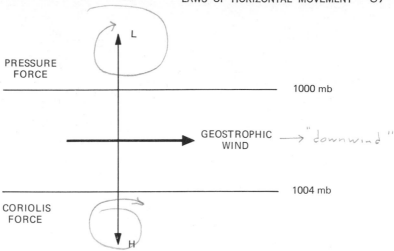

Fig. 3.3. The geostrophic wind case of balanced motion.

4. The centripetal acceleration. For a body to follow a curved path there must be an inward acceleration towards the center of rotation. This acceleration (c) is expressed by:

$$c = -\frac{mV^2}{r},$$

where m = the moving mass, V = its velocity, and r = the radius of curvature. This factor is sometimes regarded for convenience as a centrifugal force operating radially outward.[1] In the case of the earth itself this is valid. The centrifugal effect due to rotation has in fact resulted in a slight bulging of the earth's mass in low latitudes and a flattening near the poles. The decrease in apparent gravity towards the equator reflects the effect of the centrifugal force working against the gravitational attraction directed towards the earth's center. It is only necessary, therefore, to consider the forces involved in the rotation of the air about a local axis of high or low pressure. Here the curved path of the air (parallel to the isobars) is maintained by an inward-acting, or centripetal, acceleration.

Fig. 3.4 shows (for the northern hemisphere) that in a low-pressure system balanced flow is maintained in a curved path (referred to as the *gradient wind*) by the Coriolis force being weaker than the pressure force. The difference between the two gives the net centripetal acceleration inward. In the high-pressure case the inward acceleration is provided by the Coriolis

---

[1] The centrifugal force is equal in magnitude and opposite in sign to the centripetal acceleration.

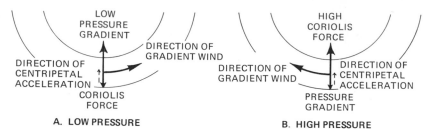

Fig. 3.4. The gradient wind case of balanced motion around a low pressure (a) and a high pressure (b).

force exceeding the pressure force. Since the pressure gradients are assumed to be equal, the different contributions of the Coriolis force in each case imply that the wind speed around the low-pressure must be less than the geostrophic value (*subgeostrophic*) whereas in the high-pressure case it is *supergeostrophic*. In reality this effect is obscured by the fact that the pressure gradient in a high is usually much weaker than in a low.

The magnitude of the centripetal acceleration is generally small, and it only becomes really important where high-velocity winds are moving in very curved paths (that is, about an intense low-pressure system). Two cases are of meteorological significance: first, in intense cyclones near the equator where the Coriolis force is negligible, and, second, in a narrow vortex such as a tornado. Under these conditions, when the large pressure gradient force provides the necessary centripetal acceleration for balanced flow parallel to the isobars, the motion is called *cyclostrophic*.

The above arguments all assume steady conditions of balanced flow. This simplification is useful but it must be noticed that whenever a pressure system is moving, or the isobar pattern diverges or converges, the air becomes subject to accelerations which cause some degree of cross-isobaric flow, indeed, pressure change depends on air displacement through the breakdown of the balanced state. If air movement were purely geostrophic there would be no growth or decay of pressure systems. The acceleration of air in moving at upper levels from a region of cyclonic isobaric curvature (subgeostrophic wind) to one of anticyclonic curvature (supergeostrophic wind) causes a fall of pressure at lower levels in the atmosphere through the removal of air aloft. The significance of this fact will be discussed in Chapter 4, Section G.

5. Frictional forces. The last force which has an important effect on air movement is that due to friction with the earth's surface. If we follow our study of a geostrophic wind a little further we find that towards the surface (that is, below about 500 meters for flat terrain) friction begins to decrease its velocity below the geostrophic value. This has an effect on the deflective force which is dependent on velocity and this force, as a result, also decreases.

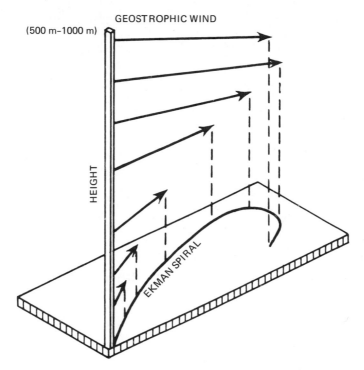

GEOSTROPHIC WIND

(500 m–1000 m)

HEIGHT

EKMAN SPIRAL

Fig. 3.5. The Ekman spiral of wind with height, in the northern hemisphere. The wind attains the geostrophic velocity at between 500 and 1000 meters in the middle and higher latitudes as frictional effects become negligible. This is a theoretical profile of wind velocity under conditions of mechanical turbulence.

As these two tendencies continue the wind consequently blows more and more obliquely across the isobars in the direction of the pressure gradient. The angle of obliqueness increases with the growing effect of friction (namely, proximity to the earth's surface) and it averages about 10°–15° at the surface over the sea and 25°–35° over land. The result is to produce a wind spiral with height (Fig. 3.5), analogous to the turning of ocean currents as the effect of surface wind stress diminishes with increasing depth. Both are referred to as *Ekman spirals*, after Ekman who investigated the variation of ocean currents with depth (see Chapter 3, Section E.3).

B. LOCAL WINDS

To the practicing meteorologist the special controls over air movement produced by local conditions often provide more problems than the effects of the major planetary forces just discussed. Diurnal tendencies are superimposed upon both the large- and small-scale patterns of wind velocity. These are particularly noticeable in the case of local winds and therefore are examined before we consider the major types of local wind regime. In normal conditions there is a general tendency for wind velocities to be least about dawn, at which time there is little vertical thermal mixing and the lower air does not therefore partake of the velocity of the more freely moving upper air (see Chapter 3, Section C). Conversely, velocities of some local winds are greatest between 1300 and 1400, for this is the time when the air suffers its greatest tendency to move vertically due to terrestrial heating,

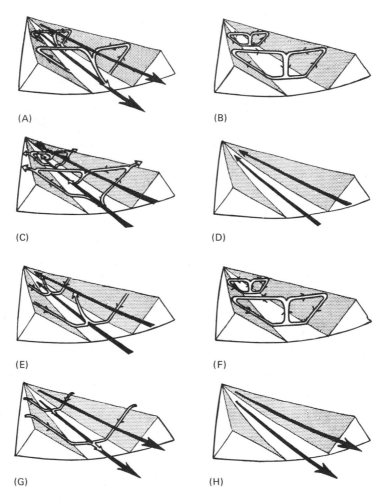

(A)　　　　　　　　　　　(B)

(C)　　　　　　　　　　　(D)

(E)　　　　　　　　　　　(F)

(G)　　　　　　　　　　　(H)

Fig. 3.6. Mountain and valley winds (from Defant, 1951). A. Sunrise. Valley cold, plains warm. B. Forenoon (0900 hrs). Valley temperature same as plains. C. Noon and early afternoon. Valley warmer than plains. D. Late afternoon. Valley continues warmer than plains. E. Evening. Valley only slightly warmer than plains. F. Early night. Valley temperature same as plains. G. Middle of night. Valley colder than plains. H. Late night. Valley much colder than plains. Note that cases B and F show only slope winds. Case D illustrates a simple valley wind and case H a simple mountain wind.

allowing it, subject to surface frictional effects, to join in the freer upper-air movement. Upper air always moves more freely than air at surface levels because it is not subject to the retarding effects of friction and obstruction.

1. Mountain and valley winds. Terrain irregularities produce special meteorological conditions of their own. During warm afternoons the laterally constricted but vertically expanding air tends to blow up the valley axis. Such winds, termed valley winds, are generally very light and require a weak regional pressure gradient in order to develop. This flow along the main valley develops simultaneously with *anabatic* (upslope) *winds* which result from a greater heating of the valley sides compared with the valley floor. At night there is a reverse process as the cold denser air at higher elevations drains into depressions and valleys; this is known as a *katabatic wind* (Fig. 3.6). The greater loss of heat by radiation affecting the highest elevations, especially if they are snow-covered, cools the immediate surface air and this eventually sinks into the valley by its own weight. Downward movements of cold air motivated in this way set up downslope winds which lead to an accumulation of cold, dense air in the valley bottom. Intense frost pockets can thus develop in mountainous, or even in moderately hilly areas. If the air drains downslope into an open valley a *mountain wind* may develop along the axis of the valley towards the plain, where it replaces warmer, less dense air. For this reason the maximum velocities of winds in mountain valleys are found to occur usually at the times of maximum diurnal cooling (for example, between 0500 and 0600 in summer).

2. Winds due to topographic barriers. Mountain ranges have important effects on the airflow across them. The displacement of air upwards over the obstacle may trigger instability if the air is conditionally unstable (see Chapter 2, Section F), whereas stable air returns to its original level in the lee of the barrier and this descent often forms the first of a series of *lee waves* (or *standing waves*) downwind as shown in Fig. 3.7. The wave form remains more or less stationary relative to the barrier with the air moving quite rapidly through it. Below the crest of the waves there may be circular air motion in a vertical plane which is termed a *rotor*. The formation of such features is naturally of vital interest to airmen. The development of lee waves is commonly disclosed by the presence of lenticular clouds, and on occasion a rotor causes reversal of the surface wind direction in the lee of high mountains. (See Plate 7.)

A related and locally very important type of wind is the föhn or Chinook. It is a strong, gusty, dry, and warm wind which develops on the lee side of a mountain range when stable air is forced to flow over the barrier by the regional pressure gradient. Often there is a loss of moisture by precipitation on the mountains (Fig. 3.8) and the air, having cooled at the saturated

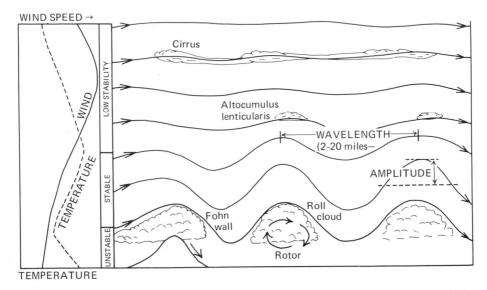

Fig. 3.7. Lee waves and rotors are produced by air flow across a long mountain range. The first wave crest usually forms less than one wavelength downwind of the ridge. There is a strong surface wind down the lee slope. Wave characteristics are determined by the wind speed and temperature relationships, shown schematically on the left of the diagram. The existence of an upper stable layer is particularly important (after Wallington, 1960).

adiabatic lapse rate above the condensation level, subsequently warms at the greater dry adiabatic lapse rate as it descends on the lee side with a consequent lowering of both the relative and absolute humidity. Recent investigations show, however, that in many instances there is no loss of moisture over the mountains and in such cases the föhn effect may be the result of wave motions forcing air from higher levels to descend. Föhn winds are common along the northern flanks of the Alps and the mountains of the

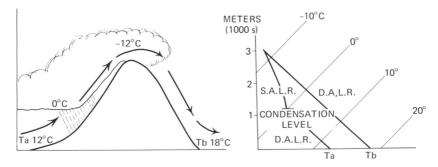

Fig. 3.8. The föhn effect when an air parcel is forced to cross a mountain range and precipitation falls. Ta refers to the temperature at the windward foot of the range and Tb to that at the leeward foot.

Caucasus and central Asia in winter and spring, when the accompanying rapid temperature rise may help to trigger-off avalanches on the snow-covered slopes. At Tashkent in central Asia, where the mean temperature in winter is about freezing point, temperatures may rise to more than 21°C during a föhn. In the same way the Chinook is a significant feature of the climate of the areas at the eastern foot of the Rockies (see Plate 6). At Pincher Creek, Alberta, a temperature rise of 21°C (38°F) occurred in 4 minutes with the onset of a Chinook on January 6, 1966. Less spectacular effects are also noticeable in the lee of the Welsh mountains, the Pennines, and the Grampians, where the importance of föhn winds lies mainly in the dispersal of cloud by the subsiding dry air.

3. Land and sea breezes. Another familiar type of air movement is the land and sea breeze (Fig. 3.9). The vertical expansion of the air column which occurs daily during the hours of heating over the more rapidly heated land (see Chapter 1, Section D.4), tilts the isobaric surfaces downwards at the coast, causing onshore winds at the surface and a compensating off-shore movement aloft. At night the air over the sea is warmer and the situation is reversed, although much of this reversal is often the effect of downslope winds blowing off the land. Fig. 3.10 shows that these local winds can have a decisive effect on coastal temperature and humidity in the tropics. The advancing cool sea air often forms a distinct line (or *front*, see Chapter 5) marked by cumulus cloud development. This often develops in summer, for example, along the Gulf Coast of Texas. On a smaller scale such features can also be observed in Britain, particularly along the south and east coasts. The sea breeze has a depth of about 1 km (3300 ft), although it thins towards the advancing edge, and may penetrate 50 km (30 miles) inland by 2100 hours. Typical velocities of such sea breezes are 4–7 m/sec (about 10–15 mph), whereas those of land breezes are usually only about 2 m/sec (about 5 mph). The counter currents aloft are usually less evident and may be obscured by the regional airflow. It is worth noting that in middle

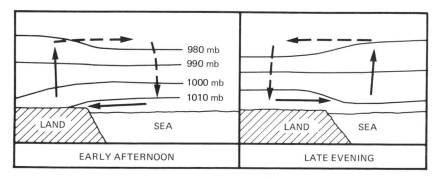

Fig. 3.9. The daytime sea breeze and the nocturnal land breeze.

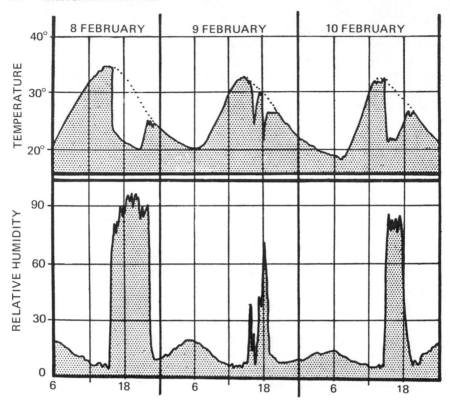

Fig. 3.10. The effect of the afternoon sea breeze on the temperature (°C) and relative humidity (%) at Joal on the Senegal coast, February 8–10, 1893 (after Angot and De Martonne. From Kuenen, 1955).

latitudes the Coriolis deflection causes turning of the onshort sea breeze (clockwise in the northern hemisphere) so that eventually it blows more or less parallel to the shore. Analogous lake breeze systems develop adjacent to large inland water bodies such as the Great Lakes.

C. VARIATION OF PRESSURE AND WIND VELOCITY WITH HEIGHT

As might be expected, changes of height reveal variations both of pressure and of wind characteristics. Study of these variations discloses some interesting facts, although explanations of these facts are as yet by no means complete. It is only possible, therefore, to outline some of the existing hypotheses which have been put forward to account for the observed characteristics.

Above the level of surface frictional effects (about 500–1000 meters) the wind increases in speed and becomes more or less geostrophic. With further height increase the reduction of air density leads to a general increase

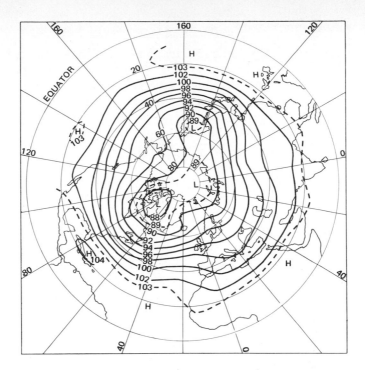

Fig. 3.11. The mean contours (in hundreds of feet) of the 700-mb pressure surface in January (above) and July (below) for 1950–9 (after O'Connor, 1961).

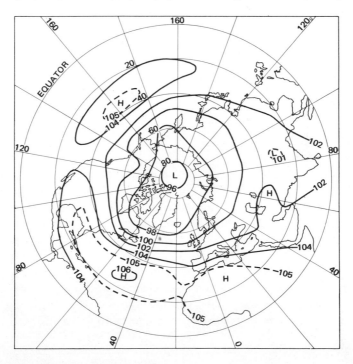

in wind speed (see Chapter 3, Section A.1). At 45°N a geostrophic wind of 14 m/sec at 3 km is equivalent to one of 10 m/sec at the surface for the same pressure gradient. There is also a seasonal variation in wind speeds aloft, these being much greater during winter months when the meridional temperature gradients are at a maximum. In addition, the persistence of these gradients tends to cause the upper winds to be more constant in direction.

1. Mean upper-air patterns. It is helpful to begin by considering the patterns of pressure and wind in the middle troposphere. These are less complicated in appearance than surface maps as a result of the diminished effects of the land masses. Rather than using pressure maps at a particular height it is convenient to depict the height of a selected pressure surface; this is termed a *contour chart* by analogy with topographic relief maps.[2] Fig. 3.11 illustrates the mean contours of the 700-mb surface for the northern hemisphere in January and July. Three features may be noted here: (1) the simplicity of the patterns, (2) the general vortex nature of the circulation, which is stronger in winter, and (3) the asymmetry of this vortex, with a primary center over the eastern Canadian Arctic and a secondary one over eastern Siberia. The major troughs and ridges form what are referred to as *long waves* (or *Rossby waves*) in the upper flow (see Chapter 4, Section F). The two major troughs at about 70°W and 150°E are thought to be induced by the combined influence on upper-air pressure and winds of large orographic barriers, like the Rockies, and heat sources such as warm ocean currents (in winter) or land masses (in summer). The subtropical high-pressure belt has only one clearly distinct cell in January over the eastern Caribbean, whereas in July cells are well developed over the Atlantic and the Pacific. In addition, the July map shows greater prominence of the subtropical high over the Sahara and southern North America. In each of these cases the subtropical cells have a pronounced east-west axis.

2. Upper winds. It is a common observation that clouds at different levels move in different directions. The wind speeds at these levels may also be markedly different, although this is not so evident to the casual observer. The gradient of wind velocity with height is referred to as the (vertical) *wind shear*, and in the free air, above the friction level, the amount of shear depends upon the temperature structure of the air. This important relationship is illustrated in Fig. 3.12, which shows that if contours are plotted of the 1000-mb and 500-mb surfaces then the gradient of the thickness of the air layer between them affects the wind velocity at the top of the layer (V_{500}). The theoretical vector wind (V_T) blowing parallel to the thickness lines (with a velocity proportional to their gradient) is termed the *thermal wind*

---

[2] The geostrophic wind concept is equally applicable to contour charts.

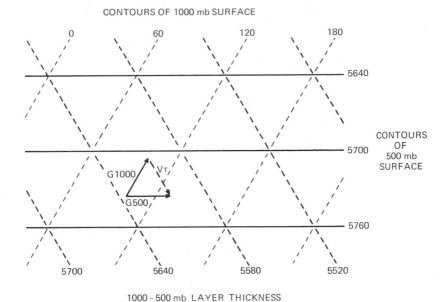

Fig. 3.12. The thermal wind and the thickness of the 1000–500-mb layer (in meters). G_{1000} is the geostrophic velocity at 1000 mb, G_{500} that at 500 mb, and V_T is the resultant 'thermal wind' blowing parallel to the thickness lines.

because the thickness of an air layer is proportional to its mean temperature—low thickness values imply cold air, high values warm air. Since the thermal wind blows with cold air (low thickness) to the left in the northern hemisphere, when viewed downwind, it is readily apparent that in the troposphere the poleward decrease of temperature should cause a large westerly component in the upper winds. Furthermore, since the meridional temperature gradient is steepest in winter the zonal westerlies are most intense at this time.

The total result of the above influences is that in the northern hemisphere most upper geostrophic winds are dominantly westerly between the subtropical high-pressure cells (centered aloft about 15°N) and the polar low-pressure center aloft. Between the subtropical high-pressure cells and the equator they are easterly. This dominant, westerly circulation reaches maximum speeds of 100–150 mph (45–67 m/sec), which even increase to 300 mph (135 m/sec) in winter. These maximum speeds are concentrated in a narrow band often situated at about 30° latitude, between 9000 and 15,000 meters, called the *jet stream*.[3] An artist's impression of one of these ribbons of high velocity is illustrated in Fig. 3.13. See also plates 10 and 11.

---

[3] The World Meteorological Organization recommends an arbitrary lower limit of 30 ms⁻¹.

This stream, which is essentially a fast-moving mass of laterally concentrated air, is in some way connected with the zone of maximum slope,
folding, or fragmentation of the tropopause, which in turn coincides with

Fig. 3.13. An artist's impression of the jet stream at about 13,000 meters (40,000 feet) over
the Midwest of the United States (after U.S. Weather Bureau. From Strahler, 1965).

the latitude of maximum poleward temperature gradient and energy transfer. The thermal wind, as described above, is a major component of the jet stream, but the basic reason for the concentration of the meridional temperature gradient in a narrow zone (or zones) is still uncertain. One theory is that the temperature gradient becomes accentuated when the upper wind pattern is confluent (see Chapter 3, Section C.5). Fig. 3.14, giving a generalized view of the wind and temperature distribution in the troposphere in winter, shows that there are two westerly jet streams. The more northerly one, termed the *Polar Front Jet Stream* (see Chapter 4, Section G), is associated with the steep temperature gradient where polar and tropical air interact, but the *Subtropical Jet Stream* is related to a temperature gradient confined to the upper troposphere. The Polar Front Jet Stream is very irregular in its latitudinal location and is commonly discontinuous, whereas the Subtropical Jet Stream is much more persistent. For these reasons the location of the mean jet stream (Fig. 3.15) primarily reflects the position of the Subtropical Jet Stream. The synoptic pattern of jet stream occurrence may be further complicated in some sectors by the presence of additional frontal zones (see Chapter 4, Section E), each associated with a jet stream. This situation is common in winter over the North American continent.

Comparison of Figs. 3.11 and 3.15 indicates that the main jet stream cores are associated with the principal troughs of the Rossby long waves. The

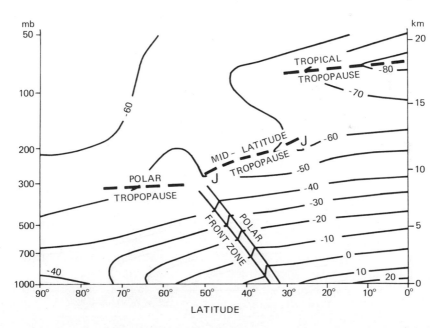

Fig. 3.14. The typical distribution of temperature and the location of the westerly jet streams (J) in the northern hemisphere in winter (after Defant and Taba, and Newton and Persson).

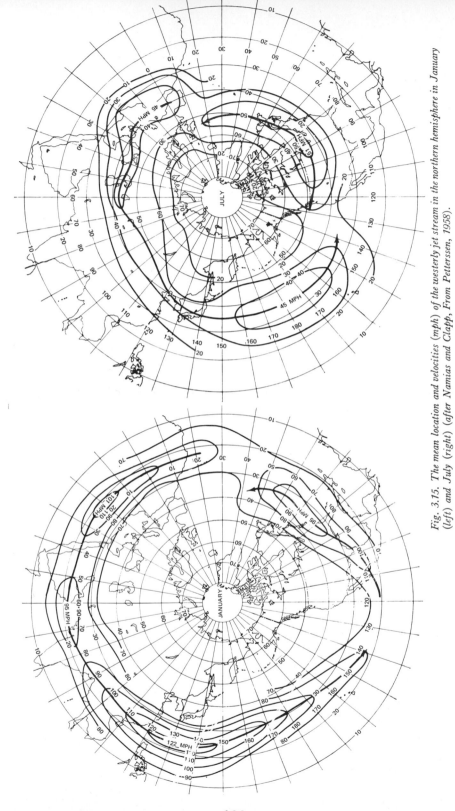

Fig. 3.15. *The mean location and velocities (mph) of the westerly jet stream in the northern hemisphere in January (left) and July (right) (after Namias and Clapp, From Petterssen, 1958).*

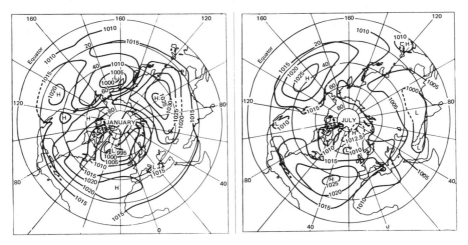

Fig. 3.16. The mean surface pressure distribution (mb) in the northern hemisphere in January (opposite) and July (above) for 1950–9 (after O'Connor, 1961).

relationships between these upper tropospheric wind systems and surface weather and climate will be considered in Chapters 4, 5, and 6.

3. Surface pressure conditions. The surface pressure patterns (Fig. 3.16) are quite complex and show considerable seasonal contrasts. Dominant are the subtropical anticyclones (high-pressure cells) suggestively situated below the Subtropical Jet Stream at about 30° latitude, extending and being thermally strengthened over the large, relatively cold continents in winter and weakening over the heated land masses in summer. In the northern hemisphere the principal subtropical high-pressure cells are located: (a) over the Bermuda-Azores ocean region (aloft the center of this cell lies over the east Caribbean); (b) over the south and southwest United States (the Great Basin or Sonoran cell)—this continental cell is naturally prone to seasonal decline, being replaced by a thermal surface low in summer; (c) over the east and north Pacific—a large and powerful cell (sometimes dividing into two, especially during the summer); and (d) over the Sahara— this, like other continental source areas, is seasonally variable both in intensity and extent, being most prominent in winter.

Thus, the subtropical high-pressure cells are the most permanent features of the surface pressure arrangement, especially over the oceans. Equatorward of these cells is the thermal low-pressure belt associated with the zone of maximum insolation and migrating with it, especially towards the heated continental interiors of the summer hemisphere. Poleward of the subtropical anticyclones lies a general zone of lower pressure, which is accentuated over the Aleutians and Iceland and over the large continents in summer. It is commonly stated that in high latitudes there is a surface high pressure resulting from the cold polar conditions, but it must be emphasized that only in spring and summer is there a true, though weak, polar high. In winter the Polar Basin is commonly an area of low pressure with the major cold-air anticyclones over Siberia and, to a lesser extent, northwestern Canada.

It is important at this point to differentiate between mean pressure patterns and the highs and lows shown on synoptic weather maps. The synoptic map is one which shows the principal pressure systems over a very large area at a given time; local wind features, for example, are ignored. The lows over Iceland and the Aleutians (Fig. 3.16) shown on mean pressure maps represent the frequent passage of deep depressions across these areas. The mean high-pressure areas, however, relate to more or less permanent highs. The intermediate areas, such as the zone about 50°–55°N, affected by traveling depressions and ridges of high pressure, appear on the mean maps as being of neither markedly high nor markedly low pressure. The movement of depressions is considered in Chapter 4, Section E (Fig. 4.14).

On comparing the surface and tropospheric pressure distributions for January (Figs. 3.11 and 3.16) it will be noticed that only the subtropical high-pressure cells extend to high levels. The reasons for this are discussed

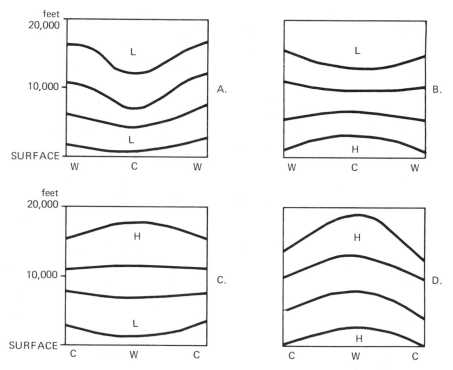

Fig. 3.17. Models of the vertical pressure distribution in cold and warm air columns. A. A surface low pressure intensifies aloft in a cold air column. B. A surface high pressure weakens aloft and may become a low pressure in a cold air column. C. A surface low pressure weakens aloft and may become a high pressure in a warm air column. D. A surface high pressure intensifies aloft in a warm air column.

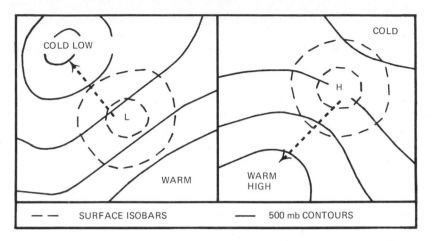

Fig. 3.18. The characteristic slope of the axes of low- and high-pressure cells with height in the northern hemisphere.

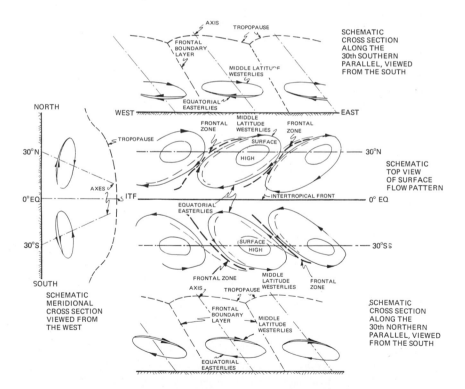

Fig. 3.19. Schematic horizontal and vertical structure of the subtropical high-pressure cells. Note particularly the convergence along the belts between the cells, the slope of the axes with height westward and equatorward, and the inclined spiral of air motion in the middle troposphere—up on the west sides (dynamically unstable air) and down on the east sides (dynamically stable air) (from Garbell, 1947).

in the next section. In summer the equatorial low-pressure belt is also evident aloft over southern Asia. The subtropical cells are still discernible at 300 mb, showing them to be a fundamental feature of the global circulation and not merely a response to surface conditions.

4. The vertical variation of pressure systems. The general relationships between surface and tropospheric pressure conditions are illustrated by the models of Fig. 3.17. A low-pressure cell at sea level with a cold core will intensify with elevation, whereas one with a warm core tends to weaken and may be replaced by high pressure. A warm air column of relatively low density causes the pressure surfaces to bulge upwards and conversely a cold, more dense air column leads to downward contraction of the pressure surfaces. Thus, a surface high-pressure cell with a cold core (a *cold anticyclone*), such as the Siberian winter anticyclone, weakens with increasing elevation and is replaced by low-pressure aloft. Cold anticyclones are shallow and rarely extend their influence above about 2500 meters (8000 ft). By contrast a surface high with a warm core (a warm anticyclone) intensifies with height. This is characteristic of the large subtropical cells which maintain their warmth through dynamic subsidence. The high surface pressure in a warm anticyclone is linked hydrostatically with cold, relatively dense air in the lower stratosphere. Conversely, a cold depression (Fig. 3.17A) is associated with a warm lower stratosphere.

Mid-latitude low-pressure cells have cold air in the rear and in consequence the axis of low pressure slopes with height towards the colder air to the west. High-pressure cells slope towards the warmest air (Fig. 3.18) and in this manner the northern hemisphere subtropical high-pressure cells are displaced 10°–15° south in latitude at the 3000-meter level, as well as towards the west (Fig. 3.19). Even so, this slope of the high-pressure axes is not constant through time and stations located between the cells may experience widely fluctuating upper winds associated with variations in the inclination of the axes.

5. Divergence, vertical motion, and vorticity. These three terms essentially hold the key to a proper understanding of modern meteorological studies of wind and pressure systems on a global scale. Mass uplift or descent of air in weather systems occurs primarily in response to dynamic factors related to horizontal airflow and is only secondarily affected by air-mass stability. Hence the significance of these factors for weather processes.

Different types of horizontal flow are shown in Fig. 3.20A. When streamlines (lines of instantaneous air motion) converge (or diverge) this is termed *confluence* (or *diffluence*). Confluence causes an increase in the velocity of air particles, but no mass accumulation. Convergence occurs when there is a net accumulation of air in a limited sector and divergence when there is net outflow. Confluence may reinforce mass convergence, but sometimes the isotach (line of equal wind speed) pattern cancels out the effect of streamline

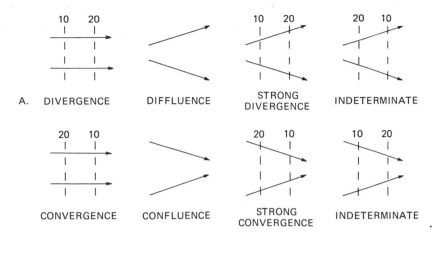

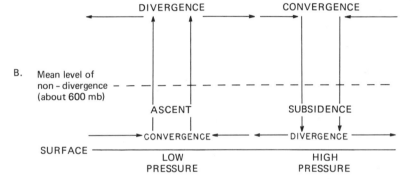

Fig. 3.20. Convergence and divergence. A. Horizontal flow patterns producing divergence and convergence. The dashed lines are schematic isopleths of wind speed (isotachs). B. The patterns of vertical motion associated with (mass) divergence and convergence in the troposphere.

confluence. If all winds were geostrophic there could be no convergence or divergence and hence no weather! Other ways in which convergence or divergence can occur are the result of surface friction effects. Onshore winds undergo convergence at low levels when the air slows down on crossing the coastline owing to the greater friction over land, whereas offshore winds accelerate and become divergent. Frictional differences can also set up coastal convergence (or divergence) if the geostrophic wind is parallel to the coastline with, for the northern hemisphere, land to the right (or left) of the air current, viewed downwind. Horizontal inflow or outflow near the surface has to be compensated by vertical motion, as illustrated in Fig. 3.20B. Air rises above a low-pressure cell and subsides over a high-pressure, with compensating divergence and convergence, respectively in the upper troposphere. In the middle troposphere there must clearly be some level at which horizontal divergence or convergence is effectively zero; the mean level of nondivergence is generally at about 600 mb. Large-scale vertical motion is extremely slow compared with convective and downdraught currents in cumulus. Typical rates in large depressions and anticyclones are of the order of 5–10 cm/sec, whereas updraughts in cumulus may exceed 10 m/sec.

Vorticity implies the rotation, or angular velocity of, minute (imaginary) particles in any fluid system. The air within a depression can be regarded as comprising an infinite number of small air parcels, each rotating cyclonically about an axis vertical to the earth's surface. Vorticity has three elements— magnitude (defined as twice the angular velocity for practical convenience), direction (the horizontal or vertical axis about which the rotation occurs), and the sense of rotation. Rotation in the same sense as the earth's rotation— cyclonic in the northern hemisphere—is defined as positive. Cyclonic vorticity may result from cyclonic curvature of the streamlines, from cyclonic shear (stronger winds on the right side of the current, viewed downwind in the northern hemisphere), or a combination of the two. Anticyclonic vorticity occurs with the corresponding anticyclonic situation. The component of vorticity about a vertical axis is referred to as the vertical vorticity. This is generally the most important, but near the ground surface frictional shear causes vorticity about an axis parallel to the surface and normal to the wind direction.

Vorticity is related not only to air motion about a cyclone or anticyclone (*relative vorticity*), but also to the location of that system on the rotating earth. The vertical component of *absolute vorticity* consists of the relative vorticity (ζ) and the latitudinal value of the Coriolis parameter, $f = 2\omega \sin \phi$ (see Chapter 3, Section A.3). At the equator the local vertical is at right angles to the earth's axis so that $f = 0$, but at the north pole cyclonic relative vorticity and the earth's rotation act in the same sense.

D. THE GLOBAL WIND BELTS

One fact which emerged from Chapter 3, Section C.3 was the importance of the subtropical high-pressure cells. Dynamic, rather than immediately thermal, in origin, and situated between 20° and 30° latitude, they seem to provide the key to the world's surface wind circulation. In the northern hemisphere the pressure gradients surrounding these cells are strongest between October and April. In terms of actual pressure, however, oceanic cells experience their highest pressure in summer, the belt being counter-balanced at low levels by thermal low-pressure conditions over the continents. Their strength and persistence clearly mark them as the dominating factor which controls the position and activities of both the trades and westerlies.

1. The trade winds. The trades (or tropical easterlies) are important because of the great extent of their activity; they blow over nearly half the globe and reach to 6–10 km at the equator over the ocean areas and even higher above the continents. They originate at low latitudes on the margins of the

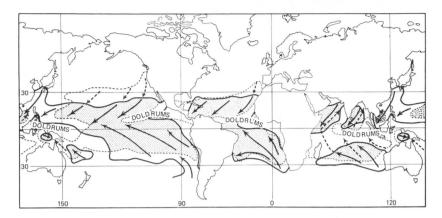

Fig. 3.21. Map of the oceanic trade wind belts and the doldrums. The limits of the trades—enclosing the area within which 50% of all winds are from the predominant quadrant—are shown by the solid (January) and dashed (July) lines. The stippled area is affected by trade-wind currents in both months. Schematic streamlines are indicated by the arrows—dashed (July) and solid (January, or both months) (based on Crowe, 1949 and 1950).

subtropical high-pressure cells, and their constancy of direction and speed is remarkable (Fig. 3.21). Trade winds, like the westerlies, are strongest during the winter half-year, which suggests that they are both controlled by the same fundamental mechanism.

The two trade-wind systems tend to converge in the *Equatorial Trough* (of low pressure). Over the oceans, particularly the central Pacific, the convergence of these air streams is pronounced and in this sector the term *Inter-Tropical Convergence Zone* (ITCZ) is applicable. Elsewhere the convergence is by no means continuous in space or time (Plate 12). Equatorward of the main *root zones* of the trades over the eastern Pacific and eastern Atlantic are regions of light, variable winds, known traditionally as the *doldrums* and much feared in past centuries by the crews of sailing ships. Their seasonal extent varies considerably; from July to September they spread westward into the central Pacific while in the Atlantic they extend to the coast of Brazil. A third major doldrum zone is located in the Indian Ocean and western Pacific. In March–April it stretches 16,000 km from east Africa to 180° longitude and it is again very extensive during October–December.

The position of the *Thermal Equator* or belt of highest seasonal temperatures is a direct function of solar heating (Fig. 1.12). There is an obvious link between this belt and the location of the Equatorial Trough which is determined by thermal lows, whereas the occurrence of the ITCZ, on the other hand, is to a considerable degree due to independent mechanisms associated with the dynamics of air circulation in low latitudes. For example, maximum convergence and uplift is commonly located several degrees south of the Equatorial Trough.

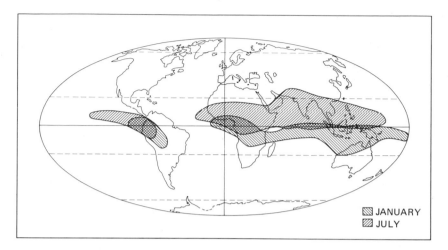

Fig. 3.22. Distribution of the equatorial westerlies in any layer below 3 km (about 10,000 ft) for January and July (after Flohn, 1960, in Indian Meteorological Department, 1960).

2. The equatorial westerlies. In the summer hemisphere, and over continental areas especially, there is a zone of generally westerly winds intervening between the two trade-wind belts (Fig. 3.22). This westerly system is well marked over Africa and southern Asia in the northern hemisphere summer when thermal heating over the continents assists the northward displacement of the Equatorial Trough (Fig. 3.21). Over Africa the westerlies reach to 2–3 km and over the Indian Ocean to 5–6 km. In the northern section of this ocean these winds are known as the *Indian Monsoon* but this is now recognized to be a complex phenomenon the cause of which is partly global and partly regional in origin (see Chapter 6, Section C). The equatorial westerlies are not simply trades of the opposite hemisphere which recurve (due to the changed direction of the Coriolis deflection) on crossing the equator, since there is *on average* a westerly component in the Indian Ocean at 2°–3°S in June and July and at 2°–3°N in December and January. Over the Pacific and Atlantic Oceans the ITCZ does not shift sufficiently far from the equator to permit the development of this westerly wind belt.

3. The mid-latitude (Ferrel) westerlies. These are the winds of the mid-latitudes emanating from the poleward sides of the subtropical high-pressure cells. They are far more variable than the trades both in direction and intensity, for in these regions the path of air movement is frequently affected by cells of low and high pressure which travel generally eastwards within the basic flow (Plate 13). Also in the northern hemisphere the preponderance of land areas with their irregular relief and changing seasonal pressure patterns tend to obscure the generally westerly air flow. The Scilly Isles, lying in the southwesterlies, record 46% of winds from between southwest and northwest, but fully 29% from the opposite sector between northeast and southeast.

The westerlies of the southern hemisphere are stronger and more constant in direction than those of the northern hemisphere because the broad expanses of ocean rule out the development of stationary pressure systems (Fig. 3.23). Kerguelen Island (49°S, 70°E) has an annual frequency of 81% of winds from between southwest and northwest and the comparable figure of 75% for Macquarie Island (54°S, 159°E) shows that this pre-dominance is widespread over the southern oceans.

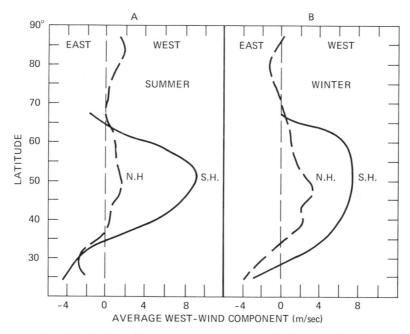

Fig. 3.23. Profiles of the average west-wind component (m/sec) at sea-level in the northern and southern hemispheres during their respective summer (A) and winter (B) seasons (after Van Loon, 1964).

4. The polar easterlies. This term is applied to winds which are supposed to occur between a polar high pressure and the belt of low pressure of the higher mid-latitudes. The polar high, as has already been pointed out, is by no means a quasi-permanent feature of the arctic circulation. Easterly winds occur mainly on the poleward sides of depressions over the North Atlantic and North Pacific and if average wind directions are calculated for entire latitude belts in high latitudes there is found to be little sign of a coherent system of polar easterlies. The situation in high latitudes of the southern hemisphere is complicated by the presence of Antarctica, but anticyclones appear to be frequent over the high plateau of eastern Antarctica and easterly winds prevail over the Indian Ocean sector of the Antarctic coastline. For example, in 1902–3 the expedition ship *Gauss* at 66°S, 90°E observed winds between northeast and southeast for 70% of the time, and at many coastal stations the constancy of easterlies may be compared with that of the trades. However, westerly components predominate over these sea areas off west Antarctica.

E. THE GENERAL CIRCULATION

The observed patterns of wind and pressure prompt consideration of the mechanisms maintaining the *general circulation* of the atmosphere—the large-scale patterns of wind and pressure which persist throughout the year or recur seasonally. Reference has already been made to one of the primary driving forces, the imbalance of radiation between lower and higher latitudes (Chapter 1, Section F.1), but it is important also to appreciate the significance

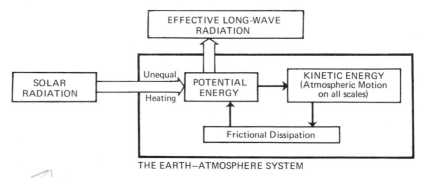

Fig. 3-24. Schematic changes of energy involving the earth-atmosphere system.

of energy transfers in the atmosphere. Energy is continually undergoing changes of form as shown schematically in Fig. 3.24. Unequal heating of the earth and its atmosphere by solar radiation generates potential energy, some of which is converted into kinetic energy by the rising of warm air and the sinking of cold air. Ultimately, the kinetic energy of atmospheric motion on all scales is dissipated by friction and small scale turbulent eddies (that is, internal viscosity). In order to maintain the general circulation, the rate of generation of kinetic energy must obviously balance its rate of dissipation. These rates are estimated to be about 2 W/square meter, which amounts only to some 1% of the average global solar radiation absorbed at the surface and in the atmosphere. In other words the atmosphere is a highly inefficient heat engine.

A second controlling factor is the angular momentum of the earth and its atmosphere. This is the tendency for the earth's atmosphere to move, with the earth, around the axis of rotation. Angular momentum is proportional to the rate of spin (that is, the angular velocity) and the square of the distance of the air parcel from the axis of rotation. With a uniformly rotating earth and atmosphere, the total angular momentum must remain constant (in other words, there is a *conservation of angular momentum*). If, therefore, a large mass of air changes its position on the earth's surface such that its distance from the axis of rotation also changes, then its angular velocity must change in a manner so as to allow the angular momentum to remain constant. Naturally angular momentum is high at the equator[4] and decreases with latitude to become zero at the pole (that is, the axis of rotation), so air moving poleward tends to acquire progressively higher eastward velocities. For example, air traveling from 42° to 46° latitude and conserving its angular momentum would increase its speed relative to the earth's surface by 29 m/sec. This is the same principle which causes an ice skater to spin more violently when her arms are progressively drawn into the body. In practice this increase of air-mass velocity is countered or masked by the other forces affecting air movement (particularly friction), but there is no doubt that many of the important features of the general atmospheric circulation result from this poleward transfer of angular momentum.

The necessity for a poleward momentum transport is readily appreciated in terms of the maintenance of the mid-latitude westerlies. These winds continually impart westerly (relative) momentum to the earth by friction and it has been estimated that they would cease altogether due to this frictional dissipation of energy in little over a week if their momentum were not continually replenished from elsewhere. In low latitudes the extensive tropical easterlies are gaining westerly (relative) momentum by friction, as a result of the earth rotating in a direction opposite to their flow, and this excess is transferred polewards with the maximum poleward transport occurring, significantly, in the vicinity of the subtropical jet stream about 200 mb at 30°N.

1. Circulations in the vertical and horizontal planes. There are two possible ways in which the atmosphere can transport heat and momentum. One is by circulation in the vertical plane as indicated in Fig. 3.25 which shows three meridional cells. The low-latitude (or Hadley) cell and its counterpart in the southern hemisphere were considered to be analogous to the convective circulations set-up when a pan of water is heated over a flame and are referred to as *thermally direct cells*. Warm air near the equator was thought to rise and generate a low-level flow towards the equator, the earth's rotation deflecting these currents which thus form the northeast and southeast trades. This explanation was put forward by Hadley in 1735, although in 1856 Ferrel pointed out that the conservation of angular momentum would be a more effective factor in causing easterlies because the Coriolis force is small in low latitudes. The low-latitude cell, according to the above scheme, would be completed by poleward counter-currents aloft with the air sinking at about 30° latitude as it is cooled by radiation. However,

---

[4] Equatorial speed of rotation is 465 m/sec.

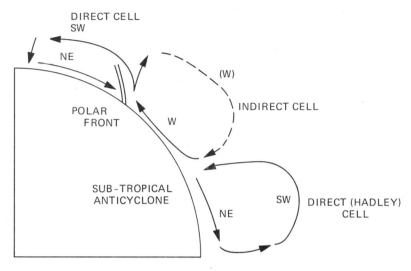

Fig. 3.25. Three-cell model of the northern hemisphere meridional circulation (after Rossby, 1941; from Barry, 1967).

this scheme is not entirely correct since the atmosphere does not have a simple heat source at the equator, the trades are not continuous around the globe (Fig. 3.21) and poleward upper flow is restricted mainly to the western ends of the subtropical high-pressure cells aloft (see Fig. 3.11).

Fig. 3.25 shows another thermally direct cell in high latitudes with cold dense air flowing out from a polar high pressure. The reality of this is doubtful, but it is in any case of limited importance to the general circulation in view of the small mass involved. It is worth noting at this point that a single direct cell in each hemisphere is not possible, because the easterly winds near the surface would slow down the earth's rotation. On average the atmosphere must rotate with the earth, requiring a balance between easterly and westerly winds over the globe.

The mid-latitude cell in Fig. 3.25 is thermally indirect and it would need to be driven by the other two. Momentum considerations indicate the necessity for upper easterlies in such a scheme, yet observations with upper-air balloons during the 1930s and 1940s demonstrated the existence of strong westerlies in the upper troposphere (Chapter 3, Section C.2). Rossby modified the 3-cell model to incorporate this fact, proposing that westerly momentum was transferred to middle latitudes from the upper branches of the cells in high and low latitudes. Such horizontal mixing could, for example, be accomplished by troughs and ridges in the upper flow.

These views underwent radical amendment from about 1948 onwards. The alternative means of transporting heat and momentum—by horizontal circulations—had been suggested in 1926 by Jeffreys but could not be tested until adequate upper-air data became available. Calculations for the northern hemisphere by Starr and White at Massachusetts Institute of Technology showed that in middle latitudes horizontal cells transport most of the required heat and momentum polewards. This operates through the mechanism of the quasi-stationary highs and the traveling highs and lows near the surface acting in conjunction with their related wave patterns aloft. The importance of such horizontal eddies for energy transport is shown in Fig. 3.26 (see also Fig. 1.25B). The modern concept of the general circulation therefore views the energy of the zonal winds as being derived from traveling waves, not from meridional circulations. In lower latitudes, however, this mechanism may be insufficient by itself to account for the total energy transport estimated to be necessary for energy balance. For such reasons the mean Hadley cell still features in current representations of the general circulation, as Fig. 3.27 shows, but the low-latitude circulation is recognized as being complex. Horizontal mixing predominates in middle and high latitudes although it is also thought that there is a weak indirect mid-latitude cell in much reduced form (Fig. 3.27). The relationship of the jet streams to regions of steep meridional temperature gradient has already been noted (see Fig. 3.14). An explanation of the two wind maxima and their role in the general circulation is lacking, but they undoubtedly form an essential part of the story.

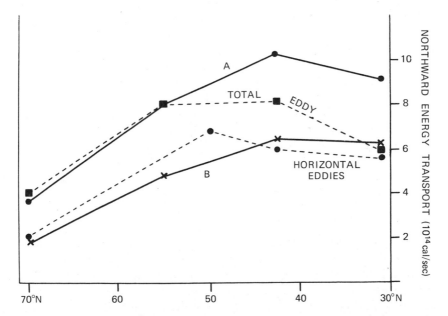

Fig. 3.26. The poleward transport of energy in the northern hemisphere. The total transport has been calculated by Gabites (A) and London (B). The contribution of the total eddy (horizontal and vertical) and of horizontal eddies alone is also shown, as calculated by Starr and White (from Tucker, 1962).

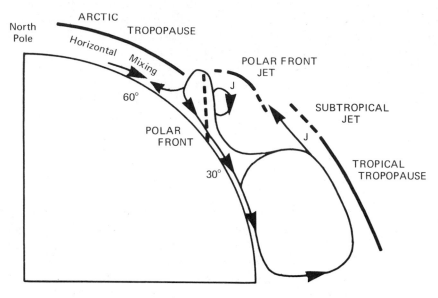

Fig. 3.27. General meridional circulation model for the northern hemisphere in winter (after Palmén, 1951; from Barry, 1967).

In the light of these theories the origin of the subtropical anticyclones which play such an important role in the world's climates may be re-examined. Their existence has been variously ascribed to the piling-up of poleward-moving air as it is increasingly deflected eastwards through the earth's rotation and the conservation of angular momentum; to the sinking of poleward currents aloft by radiational cooling; to the general necessity for high pressure near 30° latitude separating approximately equal zones of east and west winds; or to combinations of such mechanisms. An adequate theory must account not only for their permanence but also for their cellular nature and the vertical inclination of the axes. The preceding discussion shows that ideas of a simplified Hadley cell and momentum con-servation are only partially correct. It is probable that the high-level anti-cyclonic cells which are evident on *synoptic* charts (these tend to merge on mean maps) are related to anticyclonic eddies on the equatorward side of jet streams. Theoretical and observational studies show that, as a result of the latitudinal variation of the Coriolis parameter, cyclones in the westerlies tend to move poleward and anticyclonic cells equatorward. Hence the subtropical anticyclones are constantly regenerated. The cellular pattern at the surface clearly reflects the influence of heat sources. The cells are stationary and elongated north–south over the northern hemisphere oceans in summer when continental heating creates low pressure and also the meridional temperature gradient is weak. In winter, on the other hand, the zonal flow is stronger in response to a greater meridional temperature gradient and continental cooling produces east–west elongation of the cells. Undoubtedly surface and high-level factors reinforce one another in some sectors and tend to cancel out in others.

2. Variations in the circulation of the northern hemisphere. The pressure and contour patterns during certain periods of the year may be radically different from those indicated by the mean maps (see Figs. 3.11 and 3.16). These variations, of 3 to 8 weeks' duration, occur irregularly but are rather more noticeable in the winter months when the general circulation is strongest. The nature of the changes is illustrated schematically in Fig. 3.28. The zonal westerlies over middle latitudes develop waves and the troughs and ridges become accentuated, ultimately splitting up into a cellular pattern with pronounced meridional flow at certain longitudes. The strength of the westerlies between 35° and 55°N is termed the *zonal index;* strong

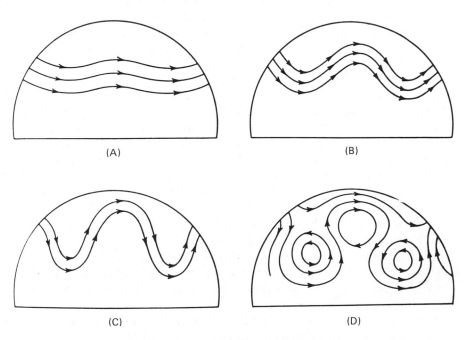

(A)

(B)

(C)

(D)

Fig. 3.28. The index cycle. A schematic illustration of the development of cellular patterns in the upper westerlies, commonly occupying 6 weeks and being especially active in February and March in the northern hemisphere. A. High zonal index. The jet stream and the westerlies lie north of their mean position. The westerlies are strong, pressure systems have a dominantly east–west orientation, and there is little north–south air-mass exchange. B and C. The jet expands and increases in velocity, undulating with increasingly larger oscillations. D. Low zonal index. Complete break-up and cellular fragmentation of the zonal westerlies. Formation of stationary deep occluding cold depressions in lower mid-latitudes and deep warm blocking anticyclones at higher latitudes. This fragmentation commonly begins in the east and extends westward at a rate of about 60° of longitude per week (after Namias. From Haltiner and Martin, 1957).

zonal westerlies are representative of high index and marked cellular patterns occur with low index. A relatively low index may also occur if the westerlies are well south of their usual latitudes and, paradoxically, such expansion of the zonal circulation pattern is associated with strong westerlies and a steep poleward temperature gradient. The cause of these variations is still uncertain although it would appear that fast zonal flow is unstable and tends to break down. This tendency is certainly increased in the northern hemisphere by the arrangement of the continents and oceans.

Detailed studies are now beginning to show that the irregular index fluctuations, together with secondary circulation features, such as cells of low and high pressure at the surface or long waves aloft, play a major role in redistributing momentum and energy. Consequently they are an integral part of the driving force behind the general circulation and are not, as used to be thought, irrelevant details superimposed upon the global wind systems. The mechanisms of the general circulation are greatly complicated by numerous interactions and feedback processes, one of the most important of which involves the oceanic circulation as outlined below. The significance of interactions between the oceanic and atmospheric heat and moisture budgets has already been discussed in Chapter 1, Section F and Chapter 2, Section A.

3. The circulation of the ocean surface. The most obvious feature of the surface oceanic circulation is the control exercised over it by the low-level planetary wind circulation, especially by the subtropical oceanic high pressure circulations and the westerlies. The oceanic circulation even partakes of the seasonal reversals of flow in the monsoonal regions of the northern Indian Ocean, off East Africa and off Northern Australia (Fig. 3.29). The Ekman effect (Chapter 3, Section A.5) causes the flow to be increasingly deflected to the right (in the northern hemisphere) and to decrease in velocity as the influence of the wind stress diminishes with depth. However, the rate of change of flow direction with depth increases with latitude, such that near the equator there are no flow reversals at depth which are characteristic of higher latitudes. The depth at which this reversal occurs decreases poleward, but averages about 50 meters over large areas of the ocean. In addition, as water moves meridionally the conservation of angular momentum implies changes in relative vorticity, with poleward-moving currents acquiring anticyclonic vorticity and equatorward-moving currents acquiring cyclonic vorticity. (See Chapter 3, Section C.5.)

Equatorward of the subtropical high pressure cells the persistent trade winds generate the broad North and South Equatorial Currents (Fig. 3.29). On the western sides of the oceans most of this water swings poleward with the airflow and thereafter increasingly comes under the influence of the Ekman deflection and of the anticyclonic vorticity effect. However, some water tends to pile up near the equator on the western sides of the oceans,

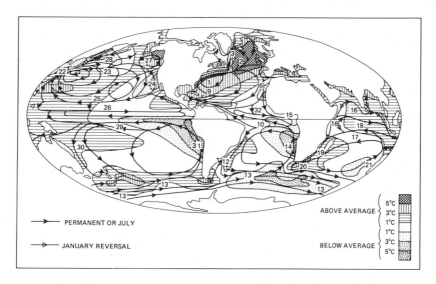

Fig. 3.29. *The general ocean current circulation of the globe, showing the mean temperature anomalies of surface ocean temperatures.*

1. *Gulf Stream.*
2. *North Atlantic Drift.*
3. *East Greenland Current.*
4. *West Greenland Current.*
5. *Labrador Current.*
6. *Canary Current.*
7. *North Equatorial Current.*
8. *Caribbean Current.*
9. *Antilles Current.*
10. *South Equatorial Current.*
11. *Brazil Current.*
12. *Falkland Current.*
13. *West Wind Drift.*
14. *Benguela Current.*
15. *Guinea Current.*
16. *Southwest and Northeast Monsoon Drift.*
17. *South Equatorial Current.*
18. *Equatorial Counter Current.*
19. *Mozambique Current.*
20. *Agulhas Current.*
21. *West Australian Current.*
22. *Kuro Shio Current.*
23. *North Pacific Drift.*
24. *California Current.*
25. *North Equatorial Current.*
26. *Equatorial Counter Current.*
27. *Alaska Current.*
28. *Kamchatka Current.*
29. *South Equatorial Current.*
30. *East Australian Current.*
31. *Peru or Humboldt Current.*
32. *Equatorial Counter Current.*

partly because here the Ekman effect is virtually absent with little poleward deflection and no reverse current at depth. To this is added some of the water which is displaced northward into the equatorial zone by the especially active subtropical high pressure circulations of the southern hemisphere. This accumulated water flows back eastward down the hydraulic gradient as compensating narrow surface Equatorial Counter Currents, unimpeded by the weak surface winds. As the circulations swing poleward round the western margins of the oceanic subtropical high pressure cells there is the tendency for water to pile up against the continents giving, for example, an appreciably higher sea level in the Gulf of Mexico than that along the Atlantic coast of the United States. This accumulated water cannot escape by sinking because of its relatively high temperature and resulting vertical stability, and it consequently continues poleward in the dominant direction of surface airflow. As a result of this movement the current gains anticyclonic vorticity which reinforces the similar tendency imparted by the winds, leading to relatively narrow currents of high velocity (for example, the Kuro Shio, Brazil, Mozambique-Agulhas and, to a less-marked extent, the East Australian Current). In the North Atlantic the configuration of the Caribbean Sea and Gulf of Mexico especially favors this pile-up of water, which is released poleward through the Florida Straits as the particularly narrow and fast Gulf Stream. These poleward currents are opposed both by their friction with the nearby continental margins and by energy losses due to turbulent diffusion, such as those accompanying the formation and cutting off of meanders in the Gulf Stream. On the poleward sides of the subtropical high-pressure cells westerly currents dominate, and where they are unimpeded by landmasses in the southern hemisphere they form the broad and swift West Wind Drift. In the northern hemisphere a great deal of the eastward-moving current in the Atlantic swings northward, leading to very anomalously high sea temperatures, and is compensated for by a southward flow of cold Arctic water at depth. However, more than half of the water mass comprising the North Atlantic Drift, and almost all that of the North Pacific Drift, swings south round the east sides of the subtropical high pressure cells, forming the Canary and California Currents. Their Southern-hemisphere equivalents are the Benguela, Humboldt or Peru, and West Australian Currents. In contrast with the currents on the west sides of the oceans, these currents acquire cyclonic vorticity which is in opposition to the anticyclonic wind tendency, leading to relatively broad flows of low velocity. In addition the deflection due to the Ekman effect causes the surface water to move westward away from the coasts, leading to upwelling of cold water from depths of 100–300 meters. Although the band of upwelling may be quite narrow (about 200 km wide for the Benguela Current) the Ekman effect spreads this cold water westward. On the poleward margins of these cold-water coasts the meridional swing of the wind belts imparts a strong seasonality to the upwelling, the California Current upwelling, for example, being particularly well marked during the period March–July.

4

Air Masses, Fronts, and Depressions

An air mass may be defined as a large body of air whose physical properties, especially temperature, moisture content and lapse rate, are more or less uniform horizontally for hundreds of kilometers. The theoretical ideal is an atmosphere where surfaces of constant pressure are not intersected by isosteric (constant-density) surfaces, so that in any vertical cross-section, as shown in Fig. 4.6, isobars and isotherms are parallel. Such an atmosphere is referred to as *barotropic*.

Three main factors tend to determine the nature and degree of uniformity of air-mass characteristics. They are: (a) the nature of the source area (from which the air mass obtains its original qualities), and the direction of movement (the physical properties of all air masses are classified according to the way in which they compare with the corresponding properties of the underlying surface region or with those of adjacent air masses); (b) changes that occur in the constitution of an air mass as it moves over long distances; and (c) the age of the air mass.

Study of the contrasting properties of different air masses leads naturally on to a consideration of air-mass boundaries or *fronts*. Their relationship to low-pressure centers and to the patterns of airflow aloft are also discussed in this chapter, and this is followed by a brief examination of the various approaches adopted in weather forecasting.

A. NATURE OF THE SOURCE AREA

We have already observed how most of the physical processes of our atmosphere result from self-regulating attempts to equalize the major differences that arise from inequalities in the world distribution of heat, moisture, and pressure. On the world scale the heat and momentum balances refer only to long-term average conditions. However, on a smaller scale, radiation and vertical mixing can produce some measure of equilibrium between the surface conditions and the properties of the overlying air mass if air remains over a given geographical region for a period of about 3 to 5 days. Naturally the chief source regions of air masses are areas of extensive, uniform surface type which are normally overlain by quasi-stationary pressure systems. These requirements are met where there is slow divergent flow from the major thermal and dynamic high-pressure cells, whereas low-pressure regions are zones of convergence into which air masses move (see Chapter 4, Section E).

Air masses are classified on the basis of two primary factors. The first is the temperature, giving arctic, polar and tropical air, and the second is the type of surface in their region of origin, giving maritime and continental categories. The major cold and warm air masses will now be discussed.

1. Cold air masses. The principal sources of cold air in the northern hemisphere are: (a) the continental anticyclones of Siberia and northern Canada, which originate continental polar (cP) air masses, and (b) the Arctic Basin, when it is dominated by high pressure (Fig. 4.1). Some classifications designate air of the latter category as continental arctic (cA), but the differ-

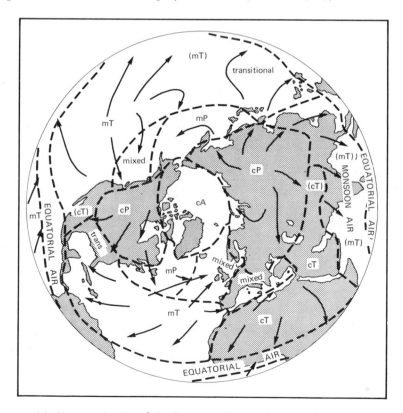

4.1. Air masses in winter (after Petterssen, 1958 and Crowe, 1965).

Table 4.1: Air-mass characteristics in winter:
(1) typical values in North America, 45–50°N (*after Godson, 1950*);
(2) monthly means over the British Isles, using surface data at Kew in place
    of the 1000 mb values (*after Belasco, 1952*);
(3) typical values in the Mediterranean (*after "Weather in the Mediterranean,"
    M.O. 391, 1962*).

T = air temperature (°C) x = humidity mixing ratio (g/kg)

Air Mass	Level (mb)			
	1000	850	700	500
cA (1) T	—	−31	−33	−42
(3) T	1	−8	−21	−36
(3) x	2.4	1.7	0.4	0.2
mA (1) T	—	−10	−21	−38
† (2) T	1	−9	−20	−40
(2) x	3.1	1.7	0.7	0.6
(3) T	4	−6	−14	−33
(3) x	4.6	2.2	1.3	0.3
cP (1) T	—	−18	−20	−33
* (2) T	−2	−12	−22	−41
(2) x	2.6	1.5	0.6	0.1
(3) T	7	−2	−13	−24
(3) x	4.5	2.6	1.3	0.4
mPw (1) T	—	5	−4	−23
** (2) T	8	1	−9	−27
(2) x	5.8	4.0	2.1	0.6
(3) T	12	2	−7	−23
(3) x	7.8	4.0	1.6	0.4
mT (1) T	—	10	0	−17
‡ (2) T	11	6	−2	−17
(2) x	6.8	5.6	3.5	1.2
(3) T	—	10	2	−14
(3) x	—	6.0	2.5	1.0
cT (3) T	—	19	5	−17
(3) x	—	1.8	1.3	0.6
Med (3) T	14	3	−3	−19
(3) x	7.0	3.7	2.5	0.9

Belasco's classification: †P₁, * A₁, ** P₇, ‡ T₁.

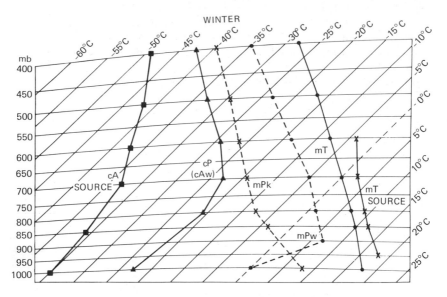

Fig. 4.2. The average vertical temperature structure for selected air masses affecting North America at about 45–50°N, recorded over their source areas or over North America in winter (after Godson, Showalter and Willett).

ences between cP and cA air masses are limited mainly to the middle and upper troposphere, where temperatures are lower in the cA air (Table 4.1).

The snow-covered source regions of these two air masses lead to marked cooling of the lower layers and, since the vapor content of cold air is very limited, the air masses generally have a mixing ratio of only 0.1–0.5 g/kg near the surface. The stability produced by the effect of surface cooling prevents vertical mixing so that further cooling occurs more slowly by radiation losses only. The effect of this radiative cooling and the tendency for air-mass subsidence in high-pressure regions combine to produce a prominent temperature inversion from the surface up to about 850 mb in typical cA and cP air. In view of their extreme dryness these air masses are characterized by small cloud amounts and only occasional light snowfalls. In summer continental heating over northern Canada and Siberia causes the virtual disappearance of their sources of cold air. The Arctic Basin source remains (see Fig. 4.3), but the cold air here is very limited in depth at this time of year.

At all seasons cA or cP air is greatly modified by a passage over the ocean. Secondary types of air mass are produced by such means and these will be considered in Chapter 4, Section B.

2. Warm air masses. These have their origins in the subtropical high-pressure cells and, during the summer season, in the great accumulations of warm surface air which characterize the heart of large land areas.

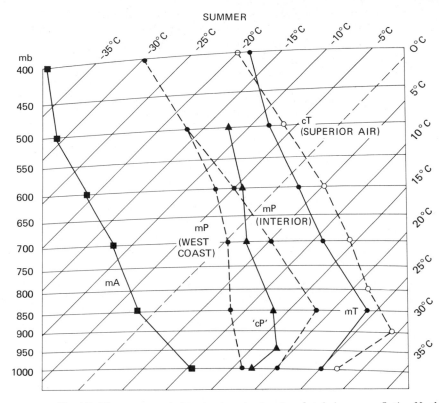

Fig. 4.3. The average vertical temperature structure for selected air masses affecting North America in summer (after Godson, Showalter and Willett).

The tropical (T) sources are either maritime (mT), originating in the oceanic subtropical high-pressure cells, or continental (cT), originating either from the continental parts of these subtropical cells (for example, as does the North African *Harmattan*) or simply associated with regions of generally light variable winds, assisted by upper tropospheric subsidence, over the major continents in summer (for example, Central Asia).

The maritime type is characterized by high temperatures (accentuated by the warming action to which the descending air is subjected), high humidity of the lower layers over the oceans, and stable stratification. Since the air is warm and moist near the surface, stratiform cloud commonly develops as the air moves polewards from its source. The continental type in winter is restricted mainly to North Africa (Fig. 4.1), where it is a warm, dry, and stable air mass. In summer warming of the lower layers by the heated land generates a steep lapse rate, but despite its instability the low relative and specific humidity prevent the development of cloud and precipitation.

The characteristics of the primary air masses are illustrated in Figs. 4.2 and 4.3 and Tables 4.1 and 4.2. In some cases their properties have been considerably affected by movement away from the source region, and it is this question which will now be considered.

Table 4.2: Air-mass characteristics in summer
(Key as for Table 4.1).

Air Mass	Level (mb)			
	1000	850	700	500
mA (1) T	—	−4	−14	−33
(2) T	14	2	−7	−25
(2) x	6.3	4.3	2.5	0.1
mP (1) T	—	11	0	−19
* (2) T	16	4	−6	−24
(2) x	8.4	3.9	2.2	0.4
(3) T	—	18	−2	−19
(3) x	—	6.0	2.5	0.8
cP (3) T	26	13	4	−14
(3) x	16.1	6.7	3.4	0.9
mT (1) T	—	18	8	−8
(2) T	19	12	4	−11
(2) x	10.8	8.1	4.5	2.4
cT (1) T	—	22	10	−11
(2) T	21	16	6	−11
(2) x	12.1	3.9	3.4	1.1
† (3) T	—	26	13	−10
(3) x	—	4.5	2.5	0.5
Med (3) T	29	19	12	−6
(3) x	14.1	7.4	3.0	0.9

Belasco's classification: * P_3.   † cT originating over Africa.

B. AIR–MASS MODIFICATION

As air masses move away from their source region they are affected by different heat and moisture exchanges with the ground surface and by dynamic processes in the atmosphere. Thus an initially barotropic air mass is gradually changed into a moderately *baroclinic* airstream in which isosteric and isobaric surfaces intersect one another. The presence of horizontal temperature gradients means that air cannot travel as a solid block maintaining an unchanging internal structure. The trajectory (that is, actual path) followed by an air parcel in the middle or upper troposphere will normally be quite different from that of a parcel nearer the surface, due to the increase of westerly wind velocity with height in the troposphere. The actual structure of an airstream at a given instant is determined to a large extent by the past history of air-mass modification processes. In spite of these qualifications the air-mass concept, as applied to the lower troposphere, nevertheless remains of considerable practical value.

1. Mechanisms of modification. The mechanisms by which air masses are modified are, for convenience, treated separately, although this rigid distinction is often not justified in practice.

(a) Thermodynamic changes. An air mass may be heated from below either by passing from a cold to a warm surface or by solar heating of the ground over which the air is located. Similarly, but in reverse, it can be cooled from below. Heating from below acts to increase air-mass instability so that the effect may be spread rapidly through a considerable thickness of air, whereas surface cooling produces a temperature inversion which greatly limits the vertical extent of the cooling. For this reason cooling mainly occurs through radiative heat loss by the air, a process which takes place only very gradually.

Changes can also occur through increased evaporation, the moisture being supplied either from the underlying surface or by precipitation from an overlying air mass layer. In reverse, the abstraction of moisture by condensation or precipitation can also cause changes. A parallel, and most important, change is the respective addition or loss of latent heat accompanying this condensation or evaporation.

(b) Dynamic changes. Dynamic (or mechanical) changes are, superficially at any rate, different from thermodynamic changes because they involve mixing or pressure changes associated with the actual movement of the air mass. The distribution of the physical properties of air masses has been shown to be considerably modified, for example, by a prolonged period of turbulent mixing (see Fig. 2.9). This process is particularly important at low levels where surface friction intensifies the natural turbulence of airflow, providing a ready mechanism for the upward transfer of the effects of thermodynamic processes.

The radiative and advective exchanges discussed previously are non-adiabatic, but the ascent or descent of air causes adiabatic changes of temperature. Large-scale lifting may result from forced ascent by a mountain barrier or from airstream convergence. Conversely, sinking may occur when high-level convergence sets up subsidence or when stable air, having been forced up over high ground by the pressure gradient, descends in its lee. Dynamic processes in the middle and upper troposphere are in fact a major cause of air-mass modification. The decrease in stability aloft, as air moves away from the areas of subsidence, is a common example of this type of mechanism.

2. The results of modification: secondary air masses. The study of the ways in which air masses change in character tells us a great deal about our weather, for many common meteorological phenomena are the product of such modification.

(a) Cold air. Continental polar air frequently streams out from Canada over the western Atlantic in winter, where it undergoes rapid transformation. Heating over the Gulf Stream Drift rapidly makes the lower layers unstable and evaporation into the air leads to sharp increases of moisture content. The turbulence associated with the convective instability is marked by gusty conditions. By the time the air has reached the central Atlantic it has become a cool, moist, maritime polar (mP) air mass. Analogous processes occur with outflow from Asia over the North Pacific (see Fig. 4.1). The weather in the airstream is typically that of bright periods and squally showers, with a variable cloud cover of cumulus and cumulonimbus. As the air moves eastward towards Europe the cooler sea surface may produce a neutral or even stable stratification near the surface, especially in summer, but subsequent heating over land will again regenerate unstable conditions. Similar conditions, but with lower temperatures (Table 4.1), result if cA air crosses sea areas in high latitudes producing maritime arctic (mA) air.

When cP air moves southward over land in winter, in central North America for example, it acquires higher temperatures and a greater tendency towards instability, but there is little gain in moisture content. The cloud type is scattered shallow cumulus which only rarely gives showers even in the afternoon when convectional instability is at a maximum. Exceptions occur in early winter around the eastern and southern shores of Hudson Bay and the Great Lakes. Until these water bodies freeze over, cold airstreams which cross them are rapidly warmed and supplied with moisture leading to locally heavy snowfalls.

Over Eurasia and North America cP air may move southwards and later recurve northwards. Some schemes of air-mass classification make allowance for such possibilities by specifying whether the air is colder (k), or warmer (w), than the surface over which it is passing. For example, cPk refers to cold, dry continental polar air which is moving over a warmer surface and is thereby likely to become unstable. Likewise, mPw indicates moist, maritime polar air which is being progressively cooled near the surface and, hence, becoming more stable.

In general, a "k" air mass has gusty, turbulent winds which make for good visibility as smoke and haze are dispersed. The instability leads to cumuliform clouds. A "w" air mass typically has stable or inversion conditions with stratiform cloud. Limited vertical mixing allows the concentration of smoke, haze, and fog at low levels. Clearly, these and similar symbols provide a convenient shorthand description of the important parameters which characterize different air masses.

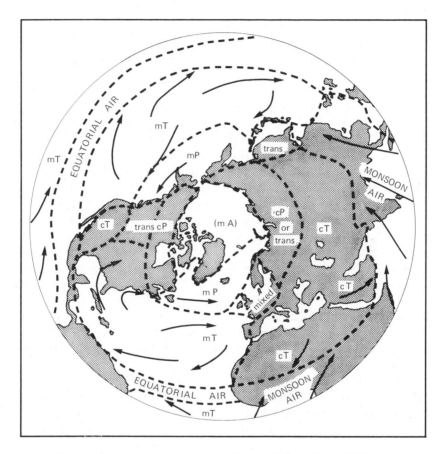

Fig. 4.4. Air masses in summer (after Petterssen, 1958 and Crowe, 1965).

Many parts of the globe must be regarded as transitional regions where the surface and air circulation produce air masses with intermediate characteristics. Northern Asia and northern Canada fall into this category in summer. In a general sense the air has affinities with continental polar air masses but these land areas, particularly the Canadian shield, have extensive bog and water surfaces so that the air is moist and cloud amounts are quite high. In a similar manner, melt-water pools and leads in the arctic pack-ice make it more appropriate to regard the area as a source of maritime arctic (mA) air in summer (Fig. 4.4).

(b) Warm air. The modification of warm air masses is usually a gradual process. Air moving poleward over progressively cooler surface becomes increasingly stable in the lower layers. In the case of mT air with high moisture content, surface cooling may produce advection fog and this is particularly common, for example, in the southwestern approaches to the English Channel during spring and early summer when the sea is still cool. If the wind velocity is sufficient for vertical mixing low stratus cloud forms in the place of fog, and drizzle may result. In addition, forced ascent of the air by high ground, or by overriding of an adjacent air mass, can produce heavy rainfall.

The cT which originates in those parts of the subtropical anticyclones situated over the arid subtropics in summer is extremely hot and dry (Table 4.2). It is typically unstable at low levels and duststorms may occur, but the dryness and the subsidence of the upper air limit cloud development. In the case of North Africa this cT air may move out over the Mediterranean rapidly acquiring moisture, with the consequent release of potential instability triggering-off showers and thunderstorm activity.

The air masses in low latitudes present considerable problems of interpretation. The temperature contrasts found in middle and high latitudes are virtually absent and what differences do exist are due principally to the presence or absence of subsidence. *Equatorial air* is usually cooler than that subsiding in the subtropical anticyclones, for example. Tropical air masses can only be differentiated meaningfully in terms of moisture content and the effects of subsidence on the lapse rate. On the equatorial sides of the subtropical anticyclones in summer the air is moving westward from areas with cool sea surfaces (for example, off northwest Africa and California) towards those of higher sea-surface temperatures. Moreover, the southwestern parts of the high-pressure cells are affected only by weak subsidence due to the vertical structure of the cells (see Chapter 3, Section C.4). As a result of these circumstances the mT air moving westwards around the equatorward sides of the subtropical highs becomes much less stable than that on the northeastern margin of the cells. Eventually such air forms the very warm, moist, unstable equatorial air of the Inter-Tropical Convergence Zone (see Figs. 4.1 and 4.4). *Monsoon air* is indicated separately in these figures, although it too may be regarded as equatorial air in summer. The special difficulties of treating tropical climatology in terms of air masses are discussed in Chapter 6.

C. THE AGE OF THE AIR MASS

Eventually the mixing and modification necessarily accompanying the movement of an air mass away from its source will cause the rate of energy exchange with the surroundings to diminish, and the various weather phenomena associated with these changes will dissipate. This process will be associated with the loss of its original identity, until finally its features merge with those of surrounding airstreams and the air may again become subject to the influence of a new source region.

Northwest Europe is shown as an area of mixed air masses in Figs. 4.1 and 4.4. This is intended to refer to the variety of sources and directions from which air may invade the region, since weather processes associated with air-mass modification and with the frontal zones separating air masses are very much in evidence. The same is also true of the Mediterranean in winter, although the area does impart its own particular characteristics to polar and other air masses which stagnate over it. Such air is termed *Mediterranean;* typical temperature and humidity values are listed in Tables 4.1 and 4.2. In winter it is convectively unstable (see Chapter 2, Section F) as a result of the moisture picked up over the Mediterranean Sea.

The length of time during which an air mass retains its original characteristics depends very much on the extent of the source area and the type of pressure pattern affecting the area. In general the lower air is changed much more rapidly than that at higher levels, although dynamic modifications aloft, which are sometimes overlooked by climatologists, are no less significant in terms of weather processes. Modern air-mass concepts must, therefore, be flexible from the point of view of both synoptic and climatological studies.

D. FRONTOGENESIS

The first real advance in our detailed understanding of mid-latitude weather variations was made with the discovery that many of the day-to-day changes are associated with the formation and movement of boundaries, or *fronts*, between different air masses. Observations of the temperature, wind directions, humidity, and other physical phenomena during unsettled periods showed that discontinuities often persist between impinging air masses of differing characteristics. The term front, for these surfaces of air-mass conflict, was a logical one to be proposed during the First World War by a group of meteorologists (including V. and J. Bjerknes, H. Solberg, and T. Bergeron) working in Norway, and their ideas are still an integral part of most weather analysis and forecasting particularly in middle and high latitudes.

1. Frontal waves. It was observed that the typical geometry of the air-mass interface, or front, resembles a wave form (see Fig. 4.5). Similar wave patterns are, in fact, found to occur on the interfaces between many different media, for example, waves on the sea surface, ripples on beach sand, aeolian sand dunes, etc. Unlike these wave forms, however, the frontal waves in the atmosphere are commonly unstable; that is, they suddenly originate, increase in size, and then gradually dissipate. For this reason the initially attractive analogy with two substances exhibiting fluid behavior, which move across or against each other and thereby generate ripples or waves on the interface,

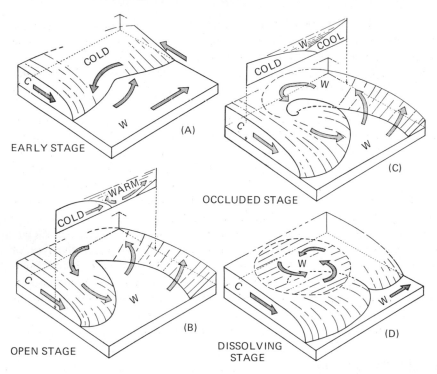

EARLY STAGE

(A)

OCCLUDED STAGE

(C)

OPEN STAGE

(B)

DISSOLVING STAGE

(D)

Fig. 4.5. Four stages in the typical development of a mid-latitude depression (from Strahler 1951).

provides an insufficient basis for explaining atmospheric wave systems. In particular, the circulation of the upper troposphere plays a key role in providing appropriate conditions for wave development and growth.

2. The characteristics of a frontal wave depression. A *depression* (also termed a low or *cyclone*[1]) is an area of relatively low pressure, with a more or less circular isobaric pattern. Such a pattern, in mid-latitudes at least, is usually associated with a convergence of contrasting air masses, between which the interface has developed into a wave form with its apex located at the center of the low-pressure area. The development of the wave form between cold and warm air traps a mass of warm air between modified cold air in front and fresh cold air in the rear. The formation of the wave also creates a distinction between the two sections of the original air-mass discontinuity for, although each section still marks the boundary between cold and warm air, the weather characteristics found in the neighborhood of each section are very different. The two sections of the frontal surface are distinguished by the names *warm front* for the leading edge of the wave and *cold front* for that of the cold air to the rear (Fig. 4.5 and Plate 14).

---

[1] This latter term is tending to be restricted to the tropical (hurricane) variety.

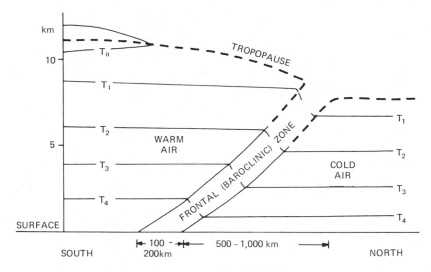

Fig. 4.6. A schematic temperature section showing barotropic air masses and a baroclinic frontal zone (assuming that density decreases with height only).

The boundary between two adjacent air masses is marked by a strongly baroclinic zone of large temperature gradient 100–200 km wide (see Chapter 4, Section B and Fig. 4.6). Sharp discontinuities of temperature, moisture, and wind properties at fronts, especially the warm front, are rather uncommon. Such discontinuities are usually the result of a pronounced surge of fresh, cold air in the rear sector of a depression, but in the middle and upper troposphere they are often caused by subsidence and may not coincide with the location of the baroclinic zone. As Fig. 4.6 shows, an upper tropospheric jet stream is closely associated with the baroclinic zone, blowing roughly parallel to the line of the upper front (see also Plate 15). This relationship is examined further in Chapter 4, Section F.

The activity of a front in terms of weather depends upon the vertical motion in the air masses. If the air in the warm sector is rising relative to the frontal zone the fronts are usually very active and are termed *ana-fronts*, whereas sinking of the warm air relative to the cold air masses gives rise to less intense *kata-fronts*.

(a) The warm front. The warm front represents the leading edge of the warm sector in the wave. The frontal zone here has a very gentle slope, of the order of $\frac{1}{2}°$–$1°$, so that the cloud systems associated with the upper portion of the front herald its approach some 12 hours or more before the arrival of the surface front. The ana-warm front, with rising warm air, has multilayered cloud which steadily thickens and lowers towards the surface position of the front. The first clouds are thin, wispy cirrus, followed by sheets of cirrus and cirrostratus, and altostratus (see Fig. 4.7). The sun is obscured

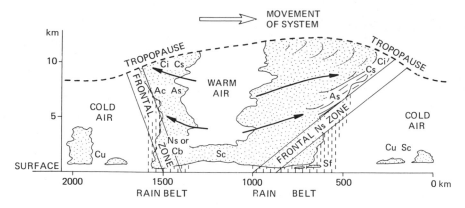

Fig. 4.7. Model of a depression with ana-fronts where the air is rising relative to each frontal surface. Note that an ana-warm front may occur with a kata-cold front (see Fig. 4.8) and vice versa (after Pedgley, 1962) (Crown Copyright reserved).

as the altostratus layer thickens and drizzle or rain begins to fall. The cloud often extends through most of the troposphere and with continuous precipitation occurring is generally designated as nimbostratus. Patches of stratus may also form in the cold air as rain falling through this air undergoes evaporation and quickly saturates it.

The descending warm air of the kata-warm front greatly restricts the development of medium- and high-level clouds. The frontal cloud is mainly stratocumulus, with a limited depth as a result of the subsidence inversions in both air masses (see Fig. 4.8). Precipitation is usually light rain or drizzle formed by coalescence since the freezing level tends to be above the inversion level, particularly in summer.

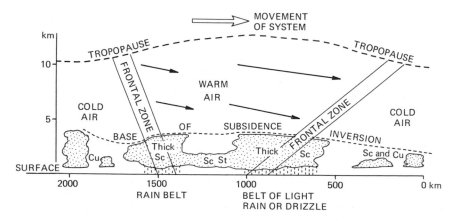

Fig. 4.8. Model of a depression with kata-fronts where the air is sinking relative to each frontal surface (after Pedgley, 1962) (Crown Copyright Reserved).

Forecasting the extent of rain belts associated with the warm front is complicated by the fact that most fronts are not ana- or kata-fronts throughout their length or even at all levels in the troposphere. For this reason, radar is being used increasingly to determine by direct means the precise extent of rain belts and even to detect differences in rainfall intensity.

At the passage of the warm front the wind veers, the temperature rises and the fall of pressure is checked. The rain becomes intermittent or ceases in the warm air and the thin stratocumulus cloud sheet may break up.

(b) The cold front. The weather conditions observed at cold fronts are equally variable, depending upon the stability of the warm sector air and the vertical motion relative to the frontal zone. The classical cold-front model is of the ana-type, and the cloud is usually cumulonimbus (Plate 16). Over the British Isles air in the warm sector is rarely unstable, so that nimbostratus occurs more frequently at the cold front (Fig. 4.7). With the kata-cold front the cloud is generally stratocumulus (Fig. 4.8) and precipitation is light. With ana-cold fronts there are usually brief, heavy downpours sometimes accompanied by thunder. The steep slope of the cold front, roughly 2°, means that the bad weather is of shorter duration than at the warm front. With the passage of the cold front the wind veers sharply, pressure begins to rise, and temperature falls. The sky may clear very abruptly, even before the passage of the surface cold front in some cases, although with kata-cold fronts the changes are altogether more gradual.

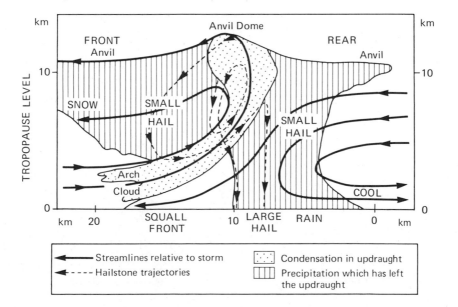

Fig. 4.9. Model of a thunderstorm which gave large hailstones. The air-flow trajectories show movement relative to the storm, which advanced from right to left (from Ludlam, 1961).

The air behind the cold front, away from the low center, commonly has an anticyclonic trajectory and hence moves at a speed greater than the geostrophic wind (see Chapter 3, Section A.4) impelling the cold front to acquire a supergeostrophic speed also. Ultimately this eliminates the warm sector, as described in the next section. Sometimes the cold air wedge advances so rapidly that in the friction layer near the surface cold air over-runs the warm air. The intrusion of this nose of cold air sets up great in-stability and the subsiding cold wedge tends to act as a scoop, forcing up the warm air where the airflow relative to this *squall-line* is contrary to the motion of the storm (see Plate 17). Such conditions generate severe frontal thunderstorms like that which struck Wokingham, Berkshire, in September 1959 (Fig. 4.9). This moved from the southwest at about 20 m/sec, steered by strong southwesterly flow aloft. The cold air subsided from high levels as a violent squall and the updraught ahead of this produced an intense hailstorm. The hailstones grow by accretion in the upper part of the up-draught, are blown ahead of the storm by strong upper winds, and begin to fall. This causes surface melting but the stone is caught up again by the advancing squall-line and reascends. The melted surface freezes, giving glazed ice as the stone is carried above the freezing level, and further growth occurs by the collection of supercooled droplets (see also Chapter 2, Sections D.3 and G).

3. The occluding stage. This stage is reached when the fresh cold air in the rear catches up with the modified cold air at the front of the depression and thus eliminates the wave form at the surface. Whenever this happens, (Fig. 4.5 and Plate 18), the wedge of warm air is pinched out at the surface and lifted bodily off the ground, as the occlusion gradually works outwards from the center of the depression along the cold front. The depression itself usually reaches its maximum degree of intensity of operation 12–24 hours after the beginning of occlusion and the lifting of the warm sector. Occlusions

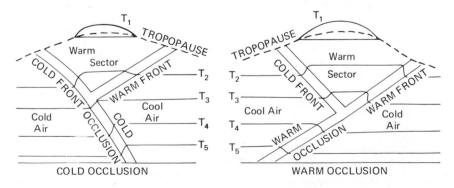

Fig. 4.10. Schematic cross-sections of a cold and a warm occlusion (after Pedgley, 1962) (Crown Copyright Reserved).

are classified as either cold or warm, the difference depending on the relative states of the cold air masses lying in front and to the rear of the warm sector (Fig. 4.10). If the air in front is colder than the air following it then the occlusion is warm, but if the reverse is so (which is more likely over the British Isles) it is termed a cold occlusion. The air in advance of the depression is most likely to be coldest when depressions occlude over Europe in winter and very cold cP air is affecting the continent.

The line of the warm air wedge aloft is associated with a zone of layered cloud (similar to that found with a warm front) and often of precipitation (see Plates 18 and 19B). Hence its position is indicated separately on some weather maps and it is referred to by Canadian meteorologists as a *trowal* (trough of warm air aloft). The passage of an occluded front and trowal brings a change back to polar air-mass weather.

The occlusion process is generally accompanied by the beginning of pressure rise at the depression center. This is much slower than the deepening of the wave in its youthful stage and may last up to six days. This time span depends primarily upon the balance of low-level convergence and upper-level divergence. The former may be accelerated by increased surface friction and the latter may decrease with changes in the pattern of upper-air flow.

The occurrence of *frontolysis* (frontal decay) is not necessarily linked with occlusion, although it represents the final phase of a front's existence. Frontal decay occurs when differences no longer exist between adjacent air masses. This may arise in four ways: through their mutual stagnation over a similar surface, as a result of both air masses moving on parallel tracks at the same speed, as a result of their movement in succession along the same track at the same speed, or by the system incorporating into itself air of the same temperature.

4. Frontal wave families. Observation has shown that frontal waves, or depressions, do not generally occur as separate units but in *families* of three or four (Fig. 4.11) with the depressions which succeed the original one forming as *secondaries* along the trailing edge of an extended cold front. Each new member follows a course which is south of its progenitor as the polar air pushes farther south to the rear of each depression in the series. Eventually the front trails far to the south and the cold polar air forms an extensive meridional wedge of high pressure terminating the sequence.

Another pattern of development may take place on the warm front, particularly at the point of occlusion, as a separate wave forms running ahead of the parent depression. This type of secondary is more likely with very cold (cA, mA, or cP) air ahead of the warm front, and its formation is encouraged when the eastward movement of the occlusion is barred by mountains. This situation often occurs when a primary depression is situated in the Davis Strait and a break-away wave forms south of Cape Farewell (the southern tip of Greenland), moving away eastwards. Analogous develop-

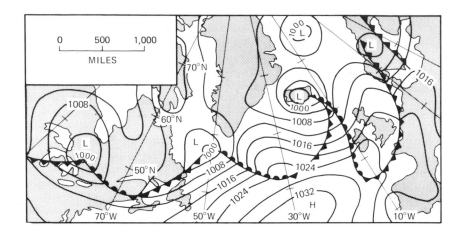

Fig. 4.11. A depression family in the North Atlantic, June 22, 1954 (after Taylor and Yates, 1958) (Crown Copyright Reserved).

ments take place in the Skagerrak–Kattegat area when the occlusion is held up by the Scandinavian mountains.

5. Mesoscale phenomena. The frontal cyclones described above are very extensive spatially and persist for several days. Intermediate in size between these and the individual cumulonimbus is a category of severe storms commonly referred to as mesoscale systems. One of these is the *squall-line*. It consists of a narrow line of thunderstorm cells which may extend for hundreds of kilometers. It is marked by a sharp veer of wind direction and very gusty conditions. The squall line often occurs ahead of a cold front maintained either as a self-propagating disturbance or by thunderstorm downdraughts. It may form a pseudo-cold front between rain-cooled air and a rainless zone within the same air mass.

Tornadoes, which may develop from such squall-line thunderstorms, are common over the Great Plains of the United States, especially in spring and early summer (Plates 13 and 20). During this period cold, dry air from the high plateaus may override maritime tropical air. Subsidence beneath the upper tropospheric westerly jet (Fig. 4.12) caps the low-level moist air forming an inversion at about 1500–2000 meters. The moist air is extended northward by a low-level southerly jet and through continuing advection the air beneath the inversion becomes progressively more warm and moist. Eventually the general convergence and ascent in the depression trigger the potential instability of the air generating large cumulus clouds which penetrate the inversion. The convective trigger is sometimes provided by the approach of the cold front towards the western edge of the moist tongue although tornadoes can occur in other synoptic situations (including high pressure areas!) if the necessary vertical contrast is present in the temperature, moisture, and wind fields. The exact tornado mechanism is still not fully understood, because of the observational problems involved. The tornado funnel has been observed to originate in the cloud base and extend towards the surface and one idea is that convergence beneath the base of cumulonimbus clouds, aided by the interaction between cold precipitation downdraughts and neighboring updraughts, may initiate rotation. Other observations suggest that the funnel forms simultaneously throughout a considerable depth of cloud, usually a towering cumulus. It appears that the upper portion of the tornado spire in this cloud may become linked to the main updraught of a neighboring cumulonimbus, thereby causing rapid removal of air from the spire and allowing a violent pressure decrease at the surface. The pressure drop is estimated to exceed 100–150 mb in some cases and it is this which makes the funnel visible by causing air entering the vortex to reach saturation. The vortex is usually only a few hundred meters in diameter (Plate 21) and in an even more restricted band around the core the winds can attain speeds of 50–200 m/sec. Destruction results not only from the high winds, for buildings near the path of the vortex may explode outwards owing to the pressure reduction outside. Tornadoes commonly occur in families and move in rather straight paths at velocities dictated by the low-level jet.

E. ZONES OF WAVE DEVELOPMENT AND FRONTOGENESIS

Fronts and associated depressions do not form everywhere and their development is restricted to well-defined areas. Frontal formation in the temperate latitudes has been studied intensively for some years and knowledge of the

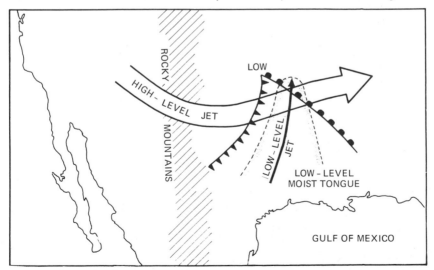

Fig. 4.12. The synoptic conditions favoring severe storms and tornadoes over the Great Plains.

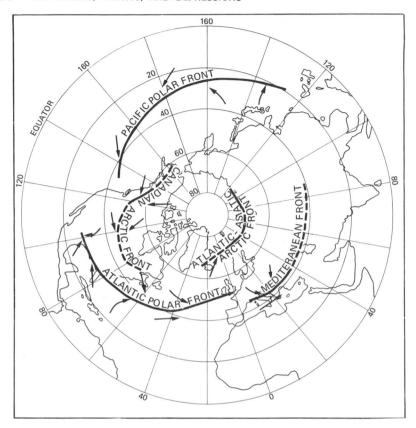

Fig. 4.13. The major northern hemisphere frontal zones in January (above) and July (opposite).

general weather conditions to be expected is reasonably accurate. Not nearly so much is known about the nature of tropical fronts but the conditions of their formation and development are unlike those normally associated with higher-latitude fronts. Increasing world air travel and the need of accurate forecasts for tropical routes is fast filling this gap. At the moment it seems that Arctic and Polar Fronts are caused primarily by gross differences in air-mass characteristics, whereas tropical discontinuities within and between somewhat similar air masses are produced mainly by the nature of the large-scale air motion and especially by confluence within an airstream or between two air currents of different humidity.

The major zones of frontal wave development are naturally those areas which are most frequently baroclinic. This is the case, for instance, off eastern Asia and eastern North America, especially in winter when there is a sharp temperature gradient between the snow-covered land and warm offshore currents. These zones are referred to respectively as the Pacific Polar and Atlantic Polar Fronts (Fig. 4.13). Their position is quite variable, but they show a general equatorward shift in winter, when the Atlantic Frontal

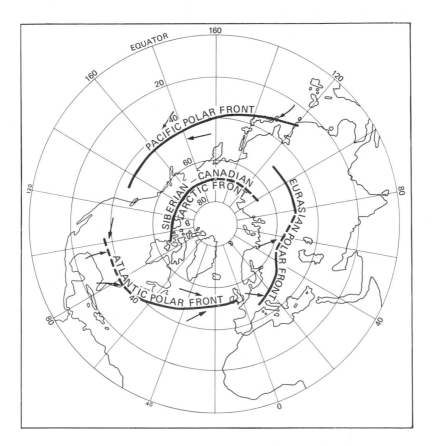

Zone may extend into the Gulf of Mexico. In this area there is convergence of air masses of different stability between adjacent subtropical high-pressure cells (this frontal zone is sometimes misleadingly termed Temperate). Depressions developing here commonly move northeastwards, sometimes following or amalgamating with others of the northern part of the Polar Front proper, or of the Canadian Arctic Front. Frontal frequency remains high across the North Atlantic, but it decreases eastward in the North Pacific, perhaps owing to a less pronounced gradient of sea-surface temperature. Frontal activity is most common in the central North Pacific when the subtropical high is split into two cells with converging air currents between.

Another section of the Polar Front, often referred to as the Mediterranean Front, is located over the Mediterranean-Caspian Sea areas in winter. At intervals, fresh Atlantic mP air, or cool cP air from southeast Europe, converges with warmer air masses, often of North African origin, over the Mediterranean Basin and initiates frontogenesis. In summer the area lies under the influence of the Azores subtropical high-pressure cell and the frontal zone is absent.

The summer locations of the Polar Front over the western Atlantic and Pacific are some 10° further north than in winter (Fig. 4.13) although the frontal zone is rather weak at this time of year. There is now a frontal zone over Eurasia and a corresponding one over middle North America. These reflect the general meridional temperature gradient and probably also the large-scale influence of orography on the general circulation (see Chapter 4, Section F).

The second major frontal zone is the Arctic Front, associated with the snow and ice margins of high latitudes. In summer this zone is developed along the Arctic coasts of Siberia and North America. In winter over North America it is formed between cA (or cP) air and Pacific maritime air modified by crossing the Coast Ranges and Rockies. There is also a less pronounced Arctic frontal zone in the North Atlantic-Norwegian Sea area, extending

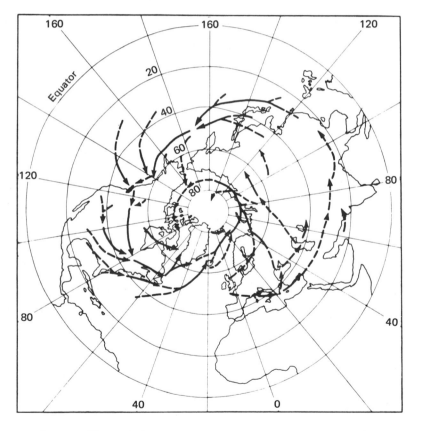

Fig. 4.14. The principal northern hemisphere depression tracks in January. The full lines show major tracks, the dashed lines secondary ones which are less frequent and less well defined. The frequency of lows is a local maximum where arrowheads end. An area of frequent cyclogenesis is indicated where a secondary track changes to a primary one or where two secondary tracks merge to form a primary (after Klein, 1957).

along the Siberian coast. A similar zone occurs all the year round in the southern hemisphere at the edge of the Antarctic ice-cap or the pack ice.

The principal tracks of depressions in the northern hemisphere in January are shown in Fig. 4.14. The major ones reflect the primary frontal zones already discussed. In summer the Mediterranean route is absent and lows move across Siberia, but the other tracks are similar although generally more zonal at this season and located in higher latitudes (around 60°N).

Between the two hemispherical belts of subtropical high pressure there is a further major world convergence zone, the Inter-Tropical Convergence Zone (or ITCZ). Formerly this was referred to as the Inter-Tropical Front (ITF), but air-mass contrasts only occur in limited sectors. This zone moves

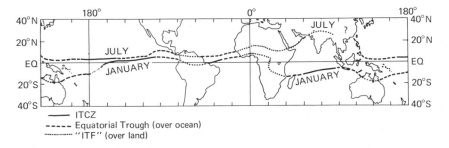

Fig. 4.15. The position of the Equatorial Trough (Inter-tropical Convergence Zone or Inter-tropical Front in some sectors) in January and July (after Crowe, 1949 and Riehl, 1954).

seasonally north and south away from the equator, as the subtropical high-pressure cell activity alternates in opposite hemispheres. The contrast between the converging air masses obviously increases with the distance of the ITCZ from the equator, and the degree of difference in their characteristics is naturally associated with considerable variation in activity along the convergence zone. Activity is most intense in June–July over southern Asia and west Africa, when the contrast between the maritime and continental air masses which are involved is at a maximum. In these sectors the zone merits the term Inter-Tropical Front, although this does not imply that it behaves like a mid-latitude frontal zone. The nature of the ITCZ and its role in tropical weather are discussed in Chapter 6. Fig. 4.15 shows the position of the Equatorial Trough (ITCZ or ITF in different sectors) in January and July.

F. SURFACE/UPPER-AIR RELATIONSHIPS AND THE FORMATION OF DEPRESSIONS

It has already been pointed out that a wave depression is associated with air-mass convergence, yet the barometric pressure at the center of the low may decrease by 10–20 mb in 12–24 hours as the system intensifies. The explanation of this apparent discrepancy is that upper air divergence removes rising air more quickly than convergence at lower levels replaces it. The superimposition of a region of upper divergence over a frontal zone is the prime motivating force of *cyclogenesis* (that is, depression formation).

The long (or *Rossby*) waves in the middle and upper troposphere, which were mentioned in Chapter 3, Section C.1, are particularly important in this respect, and it is worth considering first the reason why the hemispheric westerlies show this large-scale wave motion. The key to this problem lies in the rotation of the earth and the latitudinal variation of the Coriolis parameter (Chapter 3, Section C.5). It has been shown that for large-scale motion the (vertical) absolute vorticity $(f + \zeta)$ tends to be conserved, that is,

$$\frac{d(f + \zeta)}{dt} = 0.$$

The symbol d/dt denotes a rate of change following the motion (a total differential). Consequently, if air moves poleward so that f increases, the cyclonic vorticity tends to decrease. The curvature thus becomes anticyclonic and the current returns towards lower latitudes. If the air moves equatorward of its original latitude f decreases, requiring ζ to increase, and the resulting cyclonic curvature again deflects the current polewards. In this manner large-scale flow tends to oscillate in a wave pattern.

Rossby related the motion of these waves to their wavelength (L) and the speed of the zonal current (u). The speed of the wave (or phase speed, c) is:

$$c = u - \beta \left(\frac{L}{2\pi}\right)^2,$$

where $\beta = \partial f/\partial y$, that is, the variation of the Coriolis parameter with latitude (a local, partial differential). For stationary waves, where $c = 0$, $L = 2\pi \sqrt{u/\beta}$. At 45° latitude this stationary wavelength is 3120 km for a zonal velocity of 4 m/sec, increasing to 5400 km at 12 m/sec. The wavelengths at 60° latitude for zonal currents of 4 and 12 m/sec are, respectively, 3170 km and 6430 km. Long waves tend to remain stationary, or even to move westward against the current, (so that $c \leq 0$). Shorter waves travel eastward with a speed close to that of the zonal current and tend to be steered by the quasi-stationary long waves.

The latitudinal circumference limits the circumpolar westerly flow to between three and six major Rossby waves, and these affect the formation and movement of surface depressions. It has been pointed out that the main stationary waves tend to be located about 70°W and 150°E in response to the influence on the atmospheric circulation of orographic barriers, such as the Rocky Mountains and the Tibetan plateau, and of heat sources. On the eastern limb of troughs in the upper westerlies of the northern hemisphere the flow is normally divergent, since the gradient wind is subgeostrophic in the trough but supergeostrophic in the ridge (see Chapter 3, Section A.4). Thus the sector ahead of an upper trough is a very favorable location for a surface depression to form or deepen (see Plate 18), and it will be noted that the mean upper troughs are significantly positioned just west of the Atlantic and Pacific Polar Front Zones in winter.

With these ideas in mind we may now consider further the three-dimensional nature of depression development and the important links existing between upper and lower tropospheric flow. The basic theory relates to the vorticity equation which states that, for frictionless horizontal motion, the rate of change of the vertical component of absolute vorticity (dQ/dt or $d(f + \zeta)/dt$) is proportional to air-mass convergence ($-D$, that is, negative divergence):

$$\frac{dQ}{dt} = -DQ \quad \text{or} \quad D = -\frac{1}{Q}\frac{dQ}{dt}.$$

The conservation of vorticity equation, which we have already discussed, is in fact a special case of this relationship.

In the sector ahead of an upper trough the decreasing cyclonic vorticity causes divergence (that is, D positive), since the change in ζ outweighs that in f, thereby favoring surface convergence and low-level cyclonic vorticity. Once the surface cyclonic circulation has become established vorticity production is increased through the effects of thermal advection. Poleward transport of warm air in the warm sector and the eastward advance of the cold upper trough act to sharpen the baroclinic zone, strengthening the upper jet stream through the thermal wind mechanism (see Chapter 3, Section C.2). The vertical relationship between jet stream and front has already been shown (see Fig. 4.6) and the association as seen in plan is illustrated by the model depression sequence in Fig. 4.16. The actual relation-

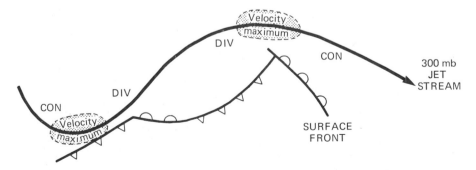

4.16. Model of the jet stream and surface fronts, showing zones of upper tropospheric divergence and convergence.

ship between the two may quite often depart from this idealized case, but the thread of maximum velocity (or *core*) of the jet is often located to the rear of the cold front.

The distribution of vertical motion upstream and downstream of this core is known to be quite different. In the area of the jet entrance (that is, upstream of the core) divergence causes lower-level air to rise on the equatorward (or right) side of the jet, whereas in the exit zone ascent is on the poleward side. It is evident that this pattern confirms the previous interpretation, with the second depression moving eastwards towards the area of maximum cyclogenetic tendency (Fig. 4.16). This pattern is of basic

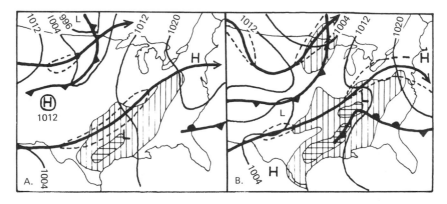

Fig. 4.17. *The relations between surface fronts and isobars, surface precipitation (0–1 inch vertical hatching; >1 inch crosshatching), and jet streams (wind speeds in excess of about 100 m.p.h. (45 m/sec) occur within the dashed lines) over the United States on (A) 20 September 1958 and (B) 21 September 1958. This illustrates how the surface precipitation area is related more to the position of the jets than to that of the surface fronts. The air over the south-central United States was close to saturation, whereas that associated with the northern jet and the maritime front was much less moist (after Richter and Dahl, 1958).*

importance in the initial deepening stage of the depression. If the upper-air pattern is unfavorable (for example, beneath left entrance and right exit zones, where there is convergence) the depression will fill. Note that to the rear of the second depression upper convergence will encourage a polar outbreak through subsidence. Fig. 4.17 shows how precipitation is often more related to the position of the jet stream than to that of surface fronts. Fig. 4.18 illustrates a typical depression family and the associated upper tropospheric jet stream (see also Plates 20 and 22).

In summary, the generation of kinetic energy which maintains a developing depression is derived primarily from the rising warm air driven by the increasing wind velocity with height (vertical shear) and the superimposition of upper tropospheric divergence over a zone of baroclinicity. Intensification of this zone further strengthens the upper winds. The upper divergence allows surface convergence and pressure fall to occur simultaneously.

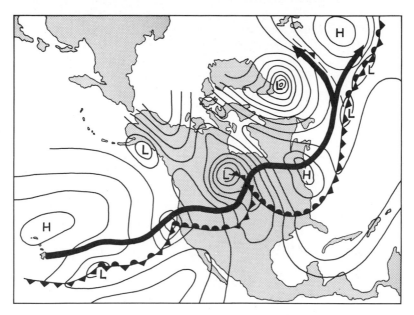

Fig. 4.18. A typical depression family and its relationship with the jet stream. The thin lines are sea-level isobars (after Vederman, 1954).

The movement of depressions is determined essentially by the upper westerlies and, as a rule of thumb, a depression center travels at about 70% of the surface geostrophic wind speed in the warm sector. Records for the United States indicate that the average speed of depressions is 32 kmph (20 mph) in summer and 48 kmph (30 mph) in winter. The higher speed in winter reflects the stronger westerly flow in response to a greater meridional temperature gradient. Shallow depressions are mainly steered by the direction of the thermal wind in the warm sector and hence their path closely follows that of the upper jet stream. Deep depressions may greatly distort the thermal pattern, however, as a result of northward transport of warm air and the southward transport of cold air. In such cases the depression usually becomes slow-moving. The movement of a depression may be additionally guided by energy sources such as a warm sea surface, which generates cyclonic vorticity, or by mountain barriers. The depression may cross obstacles, such as the Rocky Mountains or the Greenland Ice-Cap, as an upper low or trough and subsequently redevelop, aided by the lee-effects of the barrier or by fresh injections of contrasting air masses.

G. NONFRONTAL DEPRESSIONS

Not all depressions originate as frontal waves. Tropical depressions are indeed mainly nonfrontal and these are considered in Chapter 6. In middle and high latitudes four types which develop in distinctly different situations are of particular importance and interest: the lee depression, the thermal low, the polar air depression, and the cold low.

1. The lee depression. Westerly airflow which is forced over a north–south mountain barrier undergoes vertical contraction over the ridge and expansion on the lee side. This vertical movement creates compensating lateral expansion and contraction, respectively. Hence there is a tendency for divergence and anticyclonic curvature over the crest, convergence and cyclonic curvature in the lee of the barrier. Wave troughs may be set up in this way on the lee side of low hills (see Chapter 3, Section B.2) as well as major mountain chains. The airflow characteristics and the size of the barrier determine whether or not a closed low-pressure system actually develops. Such depressions, which at least initially tend to remain anchored by the barrier, are frequent in winter to the south of the Alps and the Atlas Mountains when these regions are under the influence of cold northwesterly airstreams. Fronts may occur in these depressions but it is important to recognize that the low does not form as a wave along a frontal zone.

2. The thermal low. These lows occur almost exclusively in summer, resulting from intense daytime heating of continental areas. The most impressive examples are the summer low-pressure cells over the northern part of the Indian subcontinent and over Arizona. The Iberian peninsula is another region commonly affected by such lows. The weather accompanying them is usually hot and dry, but if sufficient moisture is present the instability caused by heating may lead to showers and thunderstorms. Thermal lows normally disappear at night when the heat source is cut off, but in fact those of India and Arizona persist.

3. Polar air depressions. Polar air depressions develop mainly in winter when very unstable mP or mA air currents stream southwards along the eastern side of an extensive meridional ridge of high pressure. Such airflow may also occur in the rear of an occluding primary depression. Even when a separate low center does not form, the cold air in the rear of a primary depression may have one or more polar troughs where convergence accentuates the general cold air instability to give marked shower activity. Moreover, the gain of heat over the sea continues by day and night so that showers may develop at any time.

4. The cold low. The cold low (or *cold pool*) is usually most evident in the circulation and temperature fields of the middle troposphere. Characteristically it displays symmetrical isotherms about the low center. Surface charts may show little or no sign of these persistent systems which are frequent over northeastern North America and northeastern Siberia. They probably form as the result of strong vertical motion and adiabatic cooling in occluding baroclinic lows along the Arctic coastal margins. Such lows are especially important during the Arctic winter in that they bring large amounts of medium and high cloud which offset radiational cooling of the surface. Otherwise they usually cause no weather in the Arctic during this season. It is important to emphasize that tropospheric cold lows may be linked with either low- or high-pressure cells at the surface.

In middle latitudes cold lows may also form during periods of low-index circulation pattern (see Chapter 3, Section E.2) by the cutting-off of polar air from the main body of cold air to the north (these are sometimes referred to as *cut-off lows*). This gives rise to weather of polar air-mass type, although rather weak fronts may also be present. Such lows are commonly slow-moving and give persistent unsettled weather with thunder in summer. Heavy, early snowfalls over Colorado in October 1969 were associated with upper cold lows.

H. FORECASTING

Modern forecasting did not become possible until weather information could be rapidly collected, assembled, and processed. The main development came in the middle of the last century with the invention of wireless telegraphy, which for the first time permitted immediate analysis of weather data by the drawing of synoptic charts. These were first displayed in Britain at the Great Exhibition of 1851. Sequences of weather change were correlated with barometric pressure patterns both in space and time by such workers as Fitzroy and Abercromby, but it was not until some time later that the most helpful theoretical models of weather sequences were devised—culminating in the Bjerknes' depression model described earlier.

1. Traditional forecasting. This is carried out from the analysis of a synoptic weather map on which are plotted all the available surface weather observations for a specified time. These consist of pressure, temperature, dew point, wind, cloud, visibility, and present and past weather at each station, and they are shown by a variety of symbols or coded figures. Isobars are drawn in, usually at 4-mb intervals, and the positions of different air-mass types and their bounding fronts are inferred from the weather data. Formerly, short-term (24-hour) forecasts were based on empirical rules for the movement of surface weather systems and simple frontal concepts were used to predict depression development, but since about 1940 these methods have been greatly improved by study of the relationships between the upper-air circulation and surface systems. Now, for example, it is recognized that depressions can develop in a baroclinic zone which is not sufficiently pronounced to be termed a front.

The initial stage is to predict the future patterns of upper-air contours, surface isobars and the 1000–500 mb thickness, and then to assess what weather conditions are likely to result from the forecast situation. Inaccuracies may arise at both stages. Firstly, the network spacing and arrangement of observing stations determines much of the accuracy of the initial analysis, and since the coverage for ocean areas and for upper-air soundings is sparse only large-scale weather systems can be detected. This situation is now being remedied by satellite photography (Plates 23 and 24). Secondly, the individual depression may differ considerably from the standard models relating weather systems and weather (such as the Bjerknes' depression model). Thirdly, the small-scale nature of the turbulent motion of the atmosphere means that some weather phenomena are basically unpredictable, for example, the specific locations of shower cells in an unstable air mass. Greater precision than the "showers and bright periods" or "scattered showers" of the forecast language is impossible with present techniques. The procedure for preparing a forecast is becoming much less subjective, although in complex weather situations the skill of the experienced forecaster still makes the technique almost as much an art as a science. Detailed regional or local predictions can only be made within the framework of the general forecast situation for the country and demand thorough knowledge of possible topographic or other local effects by the forecaster.

2. Numerical forecasting. This is a much more sophisticated method which attempts to predict the physical processes in the atmosphere by means of the equations of motion originally formulated by Newton. The basic principle is that the rise or fall of surface pressure is related to mass convergence or divergence, respectively, in the overlying air column. This prediction method was proposed by L. F. Richardson who, in 1922, made a laborious test calculation which gave very unsatisfactory results. The major reason for this lack of success is the fact that the *net* convergence or divergence in an air column is a small residual term compared with the large values of convergence and divergence at different levels in the atmosphere (see Fig. 3.20). Small errors arising from observational limitations may therefore have a considerable effect on the correctness of the analysis.

Recent studies use a less direct approach. One assumes a barotropic atmosphere with geostrophic winds and hence no convergence or divergence. The movement of systems can be predicted, but not changes in intensity. Despite the great simplifications involved in the barotropic model, it has been used for forecasting 500-mb contour patterns. The latest techniques employ complex baroclinic models and include frictional and other surface effects; hence the basic mechanisms of cyclogenesis are provided for. It is noteworthy that *fields* of continuous variables, such as pressure, wind, and temperature, are handled and that fronts are regarded as secondary, derived features. The vast increase in the number of calculations which these models necessitate has required a new generation of larger faster electronic computers to allow the preparation of a forecast map to keep sufficiently ahead of the weather changes! Numerical forecasting is still in its infancy and at present such analyses are used in conjunction with others prepared by conventional methods, but the future for numerical techniques is undoubtedly one holding much promise.

3. Long-range forecasting. The methods discussed above are unsuitable for predicting the probable trend of the weather for periods of a month or more, because they are concerned with individual synoptic disturbances with a life cycle of about 3 to 7 days. Two rather different techniques will now be described, although these are by no means the only ones.

(a) Statistical methods. The United States Weather Bureau has issued 30-day forecasts twice monthly since 1948, using a method which has two principal steps. First, a mean 700-mb contour map is predicted for the coming month by a combination of several principles—extrapolation of current tendencies (such as blocking patterns) or the recognition of probable changes of regime in the large-scale atmospheric circulation, study of the possible effects of features such as snow cover and sea-surface temperature anomalies, and the examination of statistical records of the typical locations of troughs and ridges at that season. Second, the most probable anomalies of mean temperature and precipitation amount corresponding to the predicted contour map are derived from known relationships existing between them.

(b) Analog methods. Another approach to long-term forecasting developed in Britain is based on the principle that sequences of weather events may tend to follow a similar course if the initial conditions are almost identical. The problem is then to find a period with weather conditions as closely analogous as possible to the present one and to use the past sequence of events as a guide to the future. The analogs are matched from a record of patterns of monthly temperature and pressure anomalies, and of sequences of *weather types*. The latter are actually categories of the pressure pattern or direction of airflow over the country. Each category tends to be associated with a particular type of weather. The difficulty in analog prediction arises from the fact that no two patterns or sequences of weather are ever identical. There may, for example, be five reasonable analogs for a particular month, but examination of the succeeding weather sequences might show mild, rainy weather in two cases and cold spells in the other three. In the preparation of the forecast, therefore, many factors which can affect the weather trends, such as sea temperatures and the extent of snow cover, have to be taken into account. The problems facing the long-range forecasters need to be recognized before criticisms are leveled at them. The complexity of the atmosphere's behavior makes tentative wording of their predictions necessary and occasional failures inevitable at present.

A PHOTO ESSAY OF ATMOSPHERIC PHENOMENA

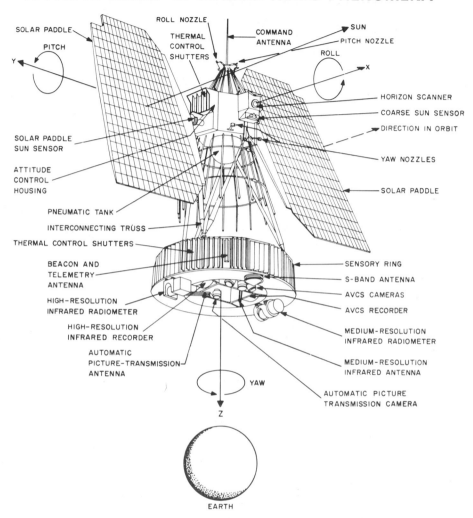

Plate 1. Nimbus II spacecraft. Launched 15 May 1966 from California into a near circular orbit with a 1180 km (733 miles) apogee and a 1094 km (680 miles) perigee (Nimbus II Users' Guide.)

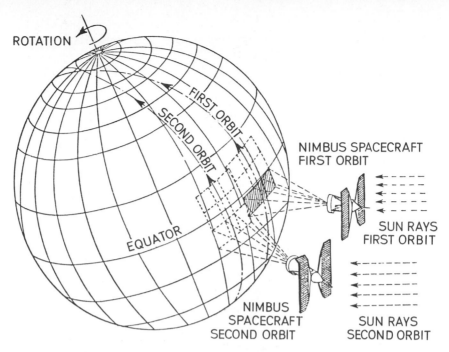

ROTATION

FIRST ORBIT

SECOND ORBIT

NIMBUS SPACECRAFT
FIRST ORBIT

SUN RAYS
FIRST ORBIT

EQUATOR

NIMBUS
SPACECRAFT
SECOND ORBIT

SUN RAYS
SECOND ORBIT

Plate 2. Nimbus II TV camera coverage of approximately 3516 × 740 km (2185 × 460 miles). Successive frames at 91-second intervals with about 20% overlap give 32 pictures for each orbit. Orbits are inclined at about 80 degrees to the equator and each successive orbit is displaced about 27 degrees in longitude (Nimbus IIP Users' Guide).

Plate 3. Cumulus orographic clouds developed over the dip-slope of the South Downs in Sussex, England. To the west, southern Hampshire is covered by stratiform clouds. This infrared photograph was taken from an elevation of about 12,000 meters (40,000 feet) (B = Burgess Hill; Br = Brighton; H = Haywards Heath; S = Shoreham; W = Worthing) (P.A.—Reuter Ltd.).

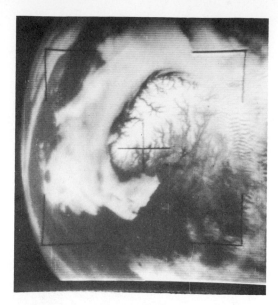

Plate 4. *The snow-covered mountains of southern Norway photographed by Tiros IX on 1 April 1965 from a height of 797 km (495 miles). To the west fog and low stratus clouds cover the sea, but near the Norwegian coast cold air draining off the mountains is sweeping the coastal waters giving clear conditions. To the east of the Scandinavian Mountains lee waves have formed in the middle troposphere with wavelengths of approximately 14 km (8.7 miles) (Environmental Science Services Administration).*

Plate 5. *View west across the north-eastern United States from Tiros VII on 9 April 1964, centered near Columbus, Ohio. The area is covered by a cold west to north-west airflow to the north of a high-pressure system centered over Arkansas. To the east of the Appalachian Mountains lee waves of some 16 km (10 miles) wavelength have formed (Environmental Science Services Administration).*

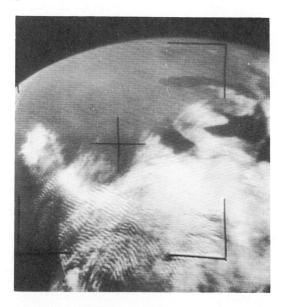

152

Plate 6. View south along the western edge of the "Chinook Arch" at Calgary, Alberta, at 14.00 hrs local time on 3 November 1965. This bank of alto-cumulus, commonly 1500 meters (5000 feet) thick with a cloud base at about 3000–3650 meters (10,000–12,000 feet) above the plains, forms on the crest of standing waves up to 80 km (50 miles) east of the foothills of the Rockies during zonal flow, generally with a pronounced inversion of temperature above polar air. This wave phenomenon occurs along a 1100 km (700 mile) belt centered on Calgary (courtesy of Ruth E. Chambers, University of Alberta).

Plate 7. View looking south-south-east from about 9000 meters (30,000 feet) along the Owen's Valley, California, showing a roll cloud developing in the lee of the Sierra Nevada mountains. The lee-wave crest is marked by the cloud layer and the vertical turbulence is causing dust to rise high into the air (W = Mount Whitney, 4418 meters (14,495 feet); I = Independence) (photograph by Robert F. Symons. Courtesy of R. S. Scorer).

154

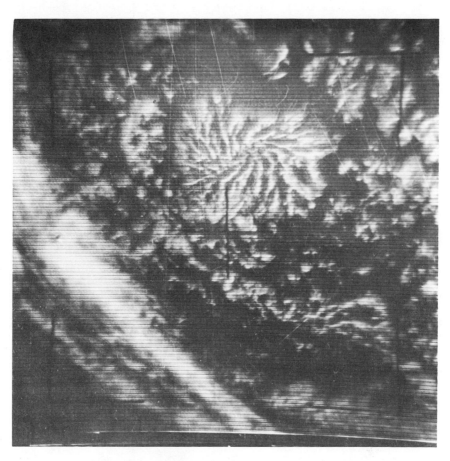

Plate 8. Radiating or dendritic cellular (actiniform) cloud pattern. These complex convective systems, some 150–250 km (90–150 miles) in diameter, were only discovered as a result of satellite photography. They usually occur in groups over areas of subsidence inversions intensified by cold ocean currents (e.g. in the low latitudes of the eastern Pacific) (Environmental Science Services Administration).

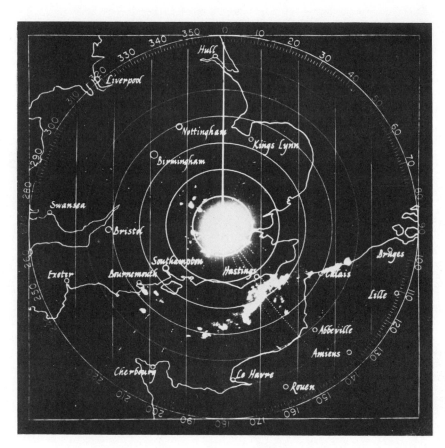

Plate 9. Photograph of Decca 45, Mk 1, 3-cm weather radar display of south-eastern England on 8 November 1955 in late afternoon. A warm occlusion had passed north-eastwards over the Straits of Dover at noon on the previous day and was followed by a mild southerly airstream in association with two centers of low pressure lying to the west of the British Isles. The unstable airstream gave temperatures well above the average, light to fresh winds, $\frac{5}{8}-\frac{7}{8}$ cloud cover and showers with scattered thunderstorms over the south-east coast. Coastlines and synoptic data have been superimposed on the radar display. (Temperatures in °F) (courtesy Decca Radar Ltd. The basic display previously appeared in "Introduction to Hydrometerology" by J. P. Bruce and R. H. Clark, Pergamon Press, London, 1966).

Plate 10. Photograph by an astronaut from Gemini XII manned spacecraft from an elevation of some 180 km (112 miles) looking south-east over Egypt and the Red Sea. The band of cirrus clouds is associated with strong upper winds, possibly concentrated as a jet stream (NASA photograph).

Plate 11. A jet-stream cloud band of cirrus and altocumulus photographed by Tiros VII over the eastern United States on 23 June 1963. At the time there was a westerly flow of cool, dry air at the surface associated with a high-pressure center over Ohio, and a belt of warm-front clouds over the Carolinas in the south-west (Environmental Science Services Administration).

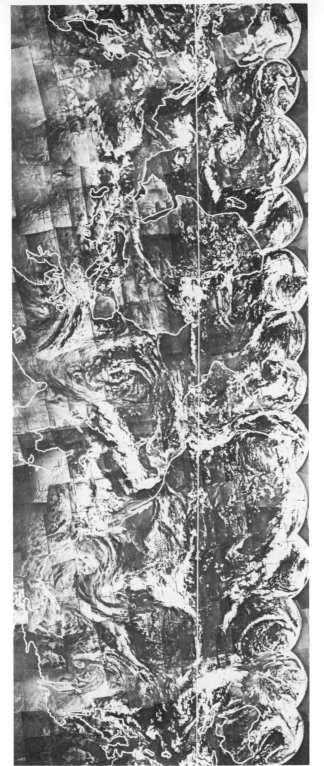

Plate 12. Mosaic of pictures from Tiros IX covering nearly the whole world taken on 13–14 February 1965. Fronts and depressions in the westerlies are especially apparent in the North Atlantic and the southern hemisphere, whereas anticyclonic clear areas are also evident (e.g. over western Spain and west of California). South of Kamchatka and north-west of the British Isles the characteristic cellular cloud patterns are associated with cold air moving south over warmer ocean surfaces behind cold fronts. The Inter-Tropical Convergence Zone cloud-belts appear just north of the equator in the Atlantic and eastern Pacific, and south of the equator in the Indian Ocean. Hurricane "Sarah" is located over the South China Sea and another hurricane over the southern Indian Ocean. North-east to south-west over northern Africa is a possible jet-stream cloud belt (U.S. Weather Bureau and Environmental Science Services Administration).

Plate 13. Photograph taken at 1446 GMT on 3 April 1968 by the Applications Technology Satellite (ATS-3) from an altitude of 22,300 miles (36,000 km) above the equator at 84°W. A number of depressions are shown over the United States (A), the North Atlantic (B), the South Atlantic (C) and the South Pacific (D), together with much cloud on the windward (i.e. eastern) side of the Central Andes (E). The depression over the United States was centered on Nebraska (A) and a cold front (F) trailing southward to the Texas coast is clearly visible. This front contained lines of thunderstorms with tops reaching to 39,000–40,000 feet (12,000–14,000 meters) and on that day 3 tornadoes and 10 funnel clouds were reported from Iowa, Oklahoma and Nebraska (from Monthly Weather Review, Vol. 96, 1968, pp. 397–398).

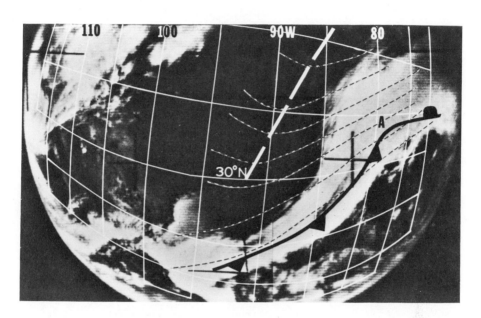

Plate 14. The initiation of a depression, centered (A) south-east of Cape Hatteras, along a frontal belt stretching north-east from the Gulf of Mexico, observed by Tiros IX on 14 February 1965. The 500-mb contour pattern is indicated by the fine dashed lines and a trough at this level (heavy dashed line) is approaching from the west (APT Users' Guide, Environmental Science Services Administration).

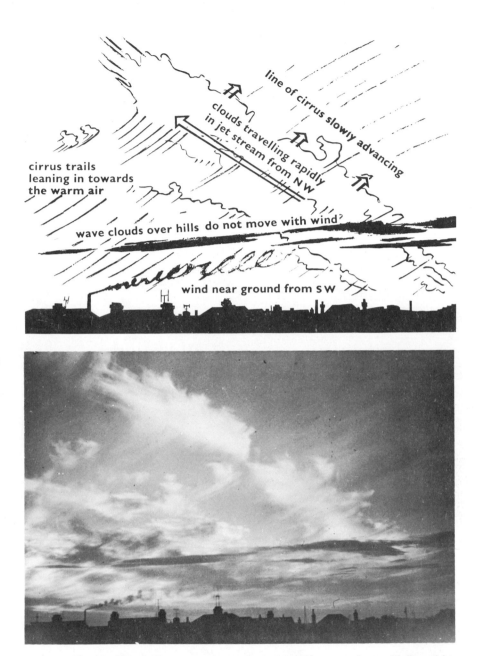

Plate 15. View looking westward towards an approaching warm front, with lines of jet-stream clouds extending from the north-west, from which trails of ice crystals are falling. In the middle levels are dark wave clouds formed in the lee of small hills by the south-westerly airflow, whereas the wind direction at the surface is more southerly—as indicated by the smoke from the chimney (photograph copyright by F. H. Ludlam. Diagram by R. S. Scorer. Both published in "Weather," Vol XVIII (8), 1963, pp. 266–7).

Plate 16. Cold front lying to the north-west of the British Isles, taken by the ESSA 6 satellite at 1130 GMT on 28 March 1968. Convective cloud is visible in the unstable air-flow behind the front, and the glitter in the North Sea is indicative of a fairly-smooth sea surface and, consequently, of light surface winds. Scandinavia is in the upper right, and snow-covered Iceland is left of center with the southern margin of sea ice lying just north-west of it (picture courtesy of Dr. P. A. Sheppard, Department of Meteorology, Imperial College, London.)

Plate 17. Thunderstorm approaching Östersund, Sweden, during late afternoon on 23 June 1955. Ahead of the region of intense precipitation there are rings of cloud formed over the squall-front (copyright F. H. Ludlam. Originally published in "Weather," Vol. XV(2), 1960, p. 63).

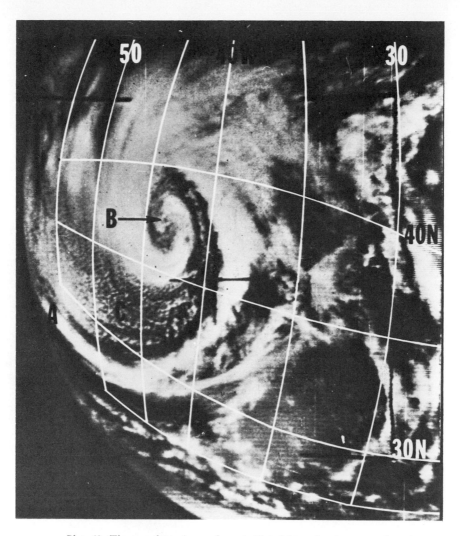

Plate 18. *The same depression as shown in Plate 14 two days later, now situated over the central North Atlantic (B), and fully occluded. The continuous cloud spiral (A–B) corresponds to the cold and occluded portions of the frontal system. In the rear of the cold front are cellular cumulus clouds (C) which are very characteristic of cold air moving over a warmer ocean surface (APT Users' Guide, Environmental Science Services Administration).*

(on the page following)

Plate 19. *The decay of a depression (Nimbus II Users' Guide, NASA photograph). A. Well-developed depression beginning to occlude at the surface and to develop a cyclonic circulation at the 500-mb level. The low clouds are mainly cumuliform. B. Fully-occluded depression. The expanded closed circulation at the surface has reached maximum intensity, as has that at the 500-mb level now centered directly above the surface vortex. The heavy overcast in the north-east quadrant is characteristic, as is the spiral pattern of low and middle-level clouds*

in the west and near the circulation center. C. Decaying depression in which the surface and 500-mb circulations are weakening and the whole organization is breaking up and exhibiting a dominantly annular, as distinct from radial, structure.

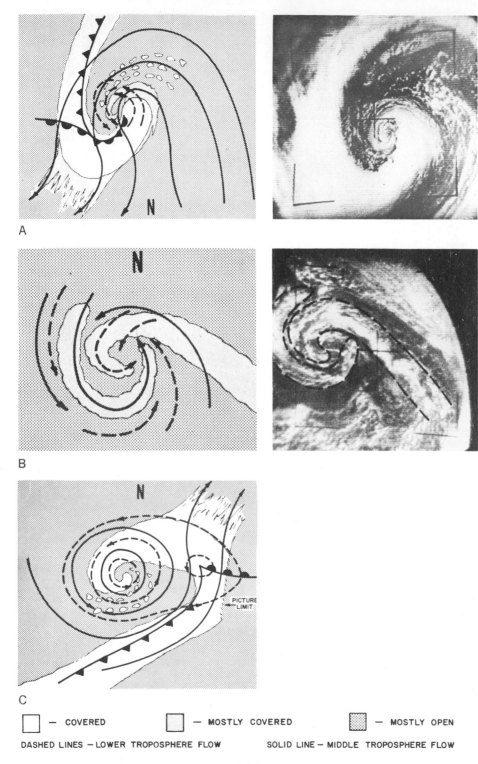

A

B

C

☐ — COVERED ▨ — MOSTLY COVERED ▨ — MOSTLY OPEN

DASHED LINES — LOWER TROPOSPHERE FLOW   SOLID LINE — MIDDLE TROPOSPHERE FLOW

Plate 20. *Oblique air photograph of a tornado taken on 15 July 1961, at 18.15 hrs local time, from an aircraft about 10 km (6 miles) north-north-east of Otis, Kansas, looking south-west (horizon tilted) from an elevation of about 600 meters (2000 feet) above the surface. This tornado occurred near the trailing edge (i.e. to the west) of a squall-line thunderstorm, which is visible with its precipitation on the far left. The top of the tornado parent cloud merged with the thunder cloud, which extended up to some 17,000 metres (55,000 feet). No lighting or rain were associated with the tornado cloud base. Little turbulence was encountered by the aircraft, but the high winds in the immediate vicinity of the base of the funnel, some 11 km (7 miles) distant, can be seen to be whipping up a dust-whirl and, in fact, drove a plank of 5 × 15 cm (2 × 6 inches) cross-section 45 cm (18 inches) into the ground on a farm some 3 km (2 miles) west of Otis (courtesy of F. C. Bates, the National Science Foundation and the University of Kansas).*

Plate 21. View from 2 miles (5 km) distance of a tornado north-east of Tracy, Minnesota, on 13 June 1968 (photograph courtesy of Eric Lantz and Associated Press). A convectively-unstable tongue of warm air extended north from Texas and by mid-afternoon its temperature had risen to 90°F (32°C), and severe thunderstorms had set in ahead of a cold front lying to its west. Surface pressure continued to drop in this belt, which was supported by a trough at 500 mb and surmounted by a jet of over 100 m.p.h. (45 m/sec) at the tropopause extending from Oregon to eastern Canada. Thunderstorm activity reached a maximum at about 1800 hrs as the unstable belt moved into Minnesota, and individual cells were shown by radar to have built up to over 50,000 feet (15,240 meters). This combination of conditions was ideal for tornado inception and on that afternoon 34 funnels were sighted within 300 miles (480 km) of Minneapolis. The Tracy tornado appeared 8 miles (13 km) south-west of the town at 1900 hours and moved north-east at 35 m.p.h. (13 m/sec) for 13 miles (21 km), cutting a 300 to 500-foot (90 to 150 meters) wide belt of total destruction through Tracy, killing 9 people, injuring 125 and causing $3 million worth of damage. Unlike most tornados, it did not lift off the ground on encountering the rough urban surface but "dug in" for its whole course until it suddenly dissolved a few seconds after the photograph was taken (description courtesy of the Director, National Severe Storms Forecast Center, Kansas City).

Plate 22. Montage of pictures taken by Nimbus II on 20 June 1966 covering almost the whole earth, with some interruptions. Africa is clearly outlined. The belts of depressions in the westerlies appear, as does the Inter-Tropical Convergence Zone in the Atlantic and eastern Pacific, and the monsoonal cloud-cover over south-east Asia (Nimbus II Users' Guide, NASA photographs).

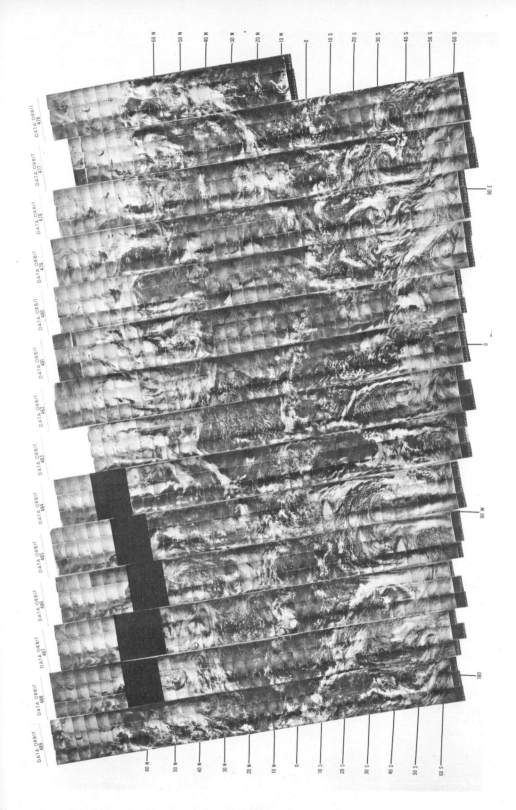

Plate 23. Tiros IV photograph of part of north-west Europe on 14 April 1962. A large frontal system is located in the Atlantic, minor troughs occur over Biscay and northern France, and cumulus clouds cover much of England (Environmental Science Services Administration photograph described by E. C. Barrett in "Geography," 1964, p. 385).

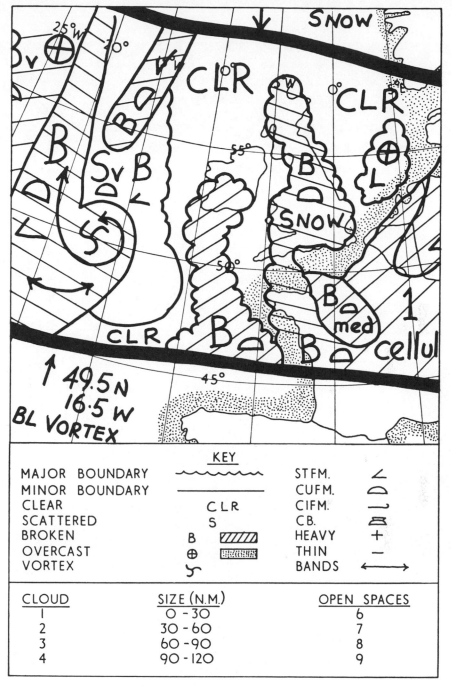

Plate 24. U.S. Weather Bureau nephanalysis for part of north-west Europe on 14 April 1962 (see Plate 23), the two heavy lines indicating the boundary of the area photographed on one orbit. This type of schematic chart is intermediate between satellite photographs and conventional weather maps (described by E. C. Barrett in "Geography," 1964, p. 385).

171

Plate 25. View south over Florida from the Gemini V manned spacecraft at an elevation of 180 km (112 miles) on 22 August 1965, with Cape Kennedy launching site in the foreground. Cumulus clouds have formed over the warmer land, with a tendency to align in east-west "streets," and are notably absent over Lake Okeechobee. In the south thunder-head anvils can be seen (NASA photograph).

Plate 26. Outflow of northerly air over India on 14 September 1966. The offshore air-stream has produced a distinct cloud band and is sweeping the tops of the convective cells over Ceylon southwards (NASA photograph).

Plate 27. An air view looking south-eastwards towards the line of high cumulus towers marking the convergence zone near the Wake Island wave trough shown in Fig. VI.5 (from Malkus and Riehl, 1964),

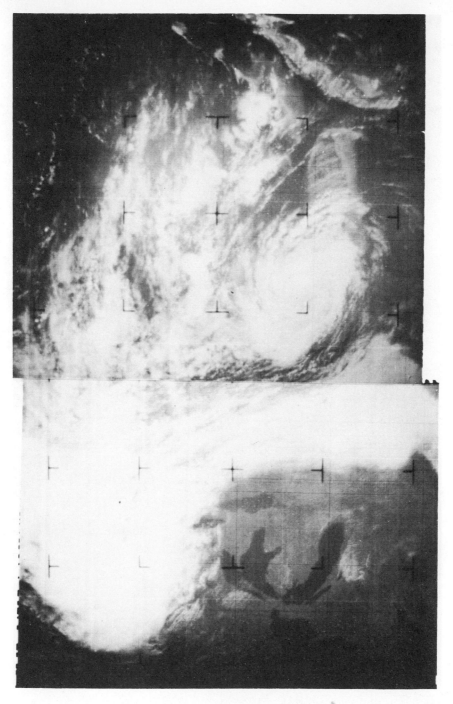

Plate 28. Montage of photographs taken by Nimbus II on 10 June 1966. The coverage stretches from Cuba to north of the Great Lakes. Hurricane "Alma" is situated north of Florida (NASA photograph).

175

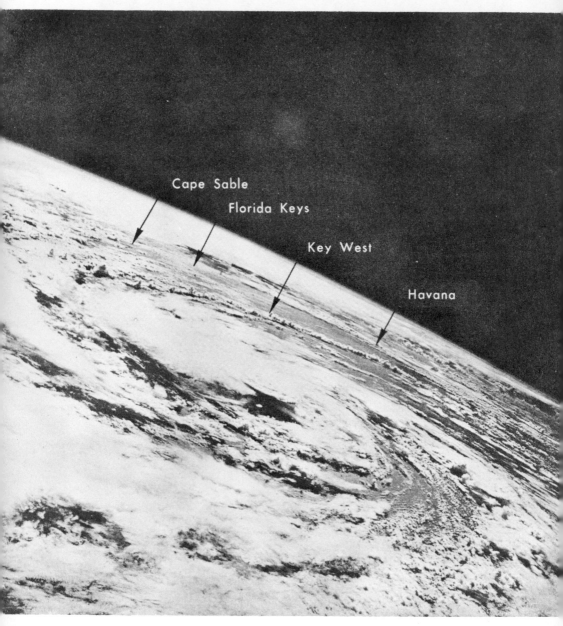

Cape Sable

Florida Keys

Key West

Havana

Plate 29. Hurricane Gladys west of Florida, photographed from Apollo 7 manned space-craft on 17 October 1968 (NASA photograph).

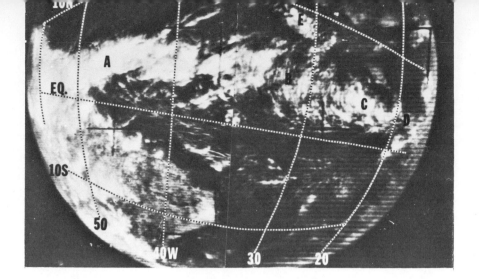

Plate 30. *Convective clouds along the Inter-Tropical Convergence Zone in the Atlantic observed by Tiros IX on 25 February 1965. The coast of Brazil appears in the south-west. Convective activity is usually very varied along this belt of about 5 degrees latitude in width, currently being especially intense at A and C and a minimum at B and D. Outside the belt (e.g. south of E) the character of the cloud changes from the Convergence-Zone cumuliform and cirriform to cellular stratocumulus (APT Users' Guide, Environmental Science Services Administration).*

Plate 31. *View over central Manhattan, New York, and the Hudson River showing skyscrapers of the Rockefeller Center (foreground) and Empire State Building (left rear) rising above the urban smoke. Several smoke sources are evident (Courtesy Aero Service, Philadelphia).*

5

Weather and Climate in Temperate Latitudes

In the two preceding chapters the general structure of the circulation and air-mass characteristics in middle latitudes have been outlined and the behavior and origin of extratropical depressions examined. The direct contribution of pressure systems to the daily and seasonal variability of weather in the westerly wind belt is quite apparent to inhabitants of the temperate lands. Nevertheless there are equally prominent contrasts of regional climate in mid-latitudes which reflect the interaction of geographical and meteorological factors. This chapter gives a selective synthesis of weather and climate in Europe and North America, drawing mainly on the principles already presented. The climatic conditions of the polar and subtropical margins of the Westerly wind belt are examined in the final sections of the chapter. As far as possible different themes are used to illustrate some of the more significant aspects of the climate in each area.

A. EUROPE

1. Pressure and wind conditions. The principal features of the mean North Atlantic pressure pattern are the Icelandic Low and the Azores High. These are present at all seasons (see Fig. 3.16), although their location and relative intensity change considerably. The upper flow in this sector undergoes little seasonal change in pattern but the westerlies decrease in strength by over half from winter to summer. The other major pressure system influencing European climates is the Siberian winter anticyclone, the occurrence of which is intensified by the extensive winter snow cover and the marked continentality of Eurasia. Atlantic depressions frequently move towards the Norwegian or Mediterranean Seas in winter, but if they travel due east they occlude and fill long before they can penetrate into the heart of Siberia. Thus the Siberian high pressure is quasipermanent at this season and when it extends westward severe conditions affect much of Europe. In summer, pressure is low over all of Asia and depressions from the Atlantic tend to follow a more zonal path. Although the depression tracks over Europe do not shift poleward in summer (as a result of the local *southward* displacement of the Atlantic Arctic Front), the depressions at this season are less intense and the diminished air-mass contrasts produce weaker fronts.

2. Oceanicity and continentality. Winter temperatures in northwest Europe are some 11°C (20°F) or more above the latitudinal average (see Fig. 1.20), a fact usually attributed solely to the presence of the North Atlantic Drift. There is, however, a complex interaction between the ocean and atmosphere. The Drift, which originates from the Gulf Stream off Florida strengthened by the Antilles Current, is primarily a wind-driven current initiated by the prevailing southwesterlies. It flows at a velocity of 16 to 32 km (10–20 miles) per day and thus, from Florida, the water takes about eight or nine months to reach Ireland and about a year to Norway (see Chapter 3, Section E.3). The southwesterly winds transport both sensible and latent heat acquired over the western Atlantic toward Europe, and although they continue to gain heat supplies over the northeastern Atlantic this local warming arises in the first place through the drag effect of the winds on the warm surface waters. Warming of air masses over the northeastern Atlantic is mainly of significance when polar or arctic air flows southeastwards from Iceland. For example, the temperature in such airstreams in winter may rise by 9°C (17°F) between Iceland and northern Scotland. By contrast, maritime tropical air cools on average about 4°C (8°F) between the Azores and southwest England in winter and summer. One very evident effect of the North Atlantic Drift is the absence of ice around the Norwegian coastline. However, as far as the climate of northwestern Europe is concerned the primary factor is the occurrence of prevailing *onshore* winds transferring heat into the area.

   The influence of maritime air masses can extend deep into Europe because there are no major topographic barriers to airflow. Hence the change to a more continental climatic regime is relatively gradual except in Scandinavia where the mountain spine produces a sharp contrast between western Norway and Sweden. There are numerous indices expressing this continentality, but most are based on the annual range of temperature. Gorczynski's continentality index (K) is:

$$K = 1.7 \frac{A}{\sin \theta} - 20.4,$$

where A is the annual temperature range (°C) and θ is the latitude angle. K ranges between about -12 at extreme oceanic stations and 100 at extreme continental stations. Some values in Europe are London 10, Berlin 21, Moscow 39. Fig. 5.1 shows the variation of this index over Europe.

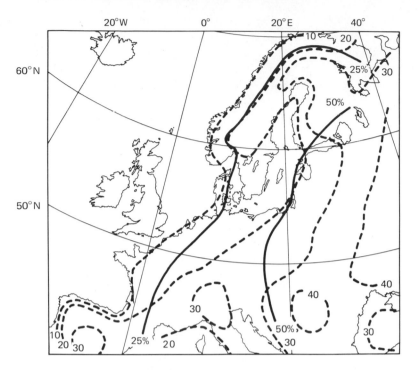

Fig. 5.1. Continentality in Europe. The indices of Gorczynski (dashed) and Berg (solid) are explained in the text.

A very different approach by Berg relates the frequency of continental air masses (C) to that of all air masses (N) as an index of continentality, or $K = C/N(\%)$. Fig. 5.1 shows that noncontinental air occurs at least half the time over Europe west of 17°E as well as over Sweden and most of Finland.

A further illustration of maritime and continental regimes is provided by Fig. 5.2. *Hythergraphs* for Valentia (Erie), Bergen, and Berlin demonstrate the seasonal changes of mean temperature and precipitation in different locations. Valentia has a winter rainfall maximum and equable temperatures as a result of oceanic situation, whereas Berlin has a considerable temperature range and a summer maximum of rainfall. Bergen receives larger rainfall totals due to orographic intensification and a maximum in autumn and winter, its temperature range being intermediate to the other two. Such averages convey only a very general impression of climatic characteristics and therefore British weather patterns will now be examined in more detail.

3. British airflow patterns and their climatic characteristics. The daily weather maps for the British Isles from 1873 to the present day have been

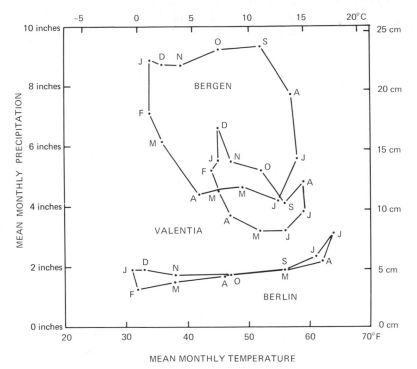

Fig. 5.2. Hythergraphs for Valentia (Eire), Bergen, and Berlin. Mean temperature and precipitation values for each month are plotted.

classified by H. H. Lamb of the Meteorological Office according to the movement of pressure systems within the general tropospheric airflow. He identifies seven major categories: Westerly (W), Northwesterly (NW), Northerly (N), Easterly (E), and Southerly (S) types—referring to the compass directions from which the systems are moving—and Cyclonic (C) or Anticyclonic (A) types when depressions or a high-pressure cell dominate the weather map.

In theory, each category should produce a characteristic type of weather, depending on the season, and many writers use the term *weather type* to convey this idea. Nevertheless, few studies have been made of the *actual* weather conditions occurring in different localities with specific patterns of airflow—a field of study known as *synoptic climatology*—hence the term *airflow type* is used here for Lamb's categories. The general weather conditions which are likely to be associated with a particular airflow type over the British Isles are summarized in Table 5.1.

Each airflow pattern may of course result in several air masses affecting an area as it is crossed by depressions traveling within the general airflow. Generalized air-mass relationships for each British airflow type are listed in Table 5.2. Such climatological characteristics of dominant airflows have

Table 5.1:   General weather characteristics of Lamb's Airflow Types.

Type

Westerly	Unsettled weather with variable wind directions as depressions cross the country. Mild and stormy in winter, generally cool and cloudy in summer.
Northwesterly	Cool, changeable conditions. Strong winds and showers affect windward coasts especially, but the southern part of Britain may have dry, bright weather.
Northerly	Cold weather at all seasons, often associated with polar lows or troughs. Snow and sleet showers in winter, especially in the north and east.
Easterly	Cold in the winter half-year, sometimes very severe weather in the south and east with snow or sleet. Warm in summer with dry weather in the west. Occasionally thundery.
Southerly	Generally warm and thundery in summer. In winter it may be associated with a depression in the Atlantic giving mild, damp weather especially in the southwest or with a high over central Europe, in which case it is cold and dry.
Cyclonic	Rainy, unsettled conditions often accompanied by gales and thunderstorms. This type may refer either to the rapid passage of depressions across the country or to the persistence of a deep depression.
Anticyclonic	Warm and dry in summer apart from occasional thunderstorms. Cold in winter with night frosts and fog especially in autumn.

Table 5.2:   Airflow types and corresponding air masses.

Westerly	mP, mPw, mT
Northwesterly	mP, mA
Northerly	mA
Easterly	cA, cP
Southerly	mT or cT (summer); mT or cP (winter)
Cyclonic	mP, mPw, mT

been quite thoroughly investigated in many countries and the following account of British synoptic climatology can be taken as a typical example of this approach to the study of climatology.

The general properties of air masses have been examined in Chapter 4, but certain aspects are of particular interest in respect of the British climate. The frequency of air masses in January, based on a study by Belasco for 1938–49, is illustrated for Kew in Fig. 5.3. There is a clear predominance of polar maritime (mP and mPw) air which has a frequency of 30% or more in all months except March. The maximum frequency of mP air at Kew is

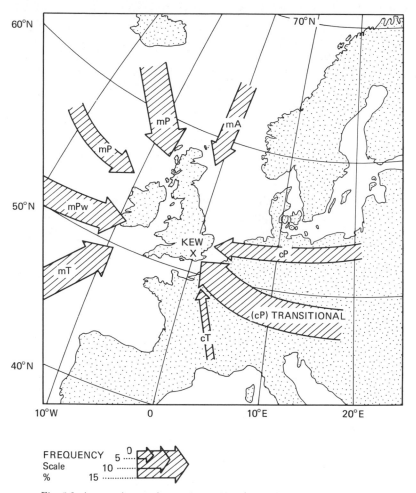

Fig. 5.3. Average air-mass frequencies for Kew (London) in January. Anticyclonic types are included according to their direction of origin (based on Belasco, 1952).

33% (with a further 10% mPw) in July. The proportion is even greater in western coastal districts with mP and mPw occurring in the Hebrides, for example, on at least 38% of days throughout the year.

Northwesterly mP airstreams produce cool, showery weather at all seasons, especially on windward coasts. The air is unstable with cumuliform cloud, although inland in winter and at night the clouds disperse, giving low night temperatures. Over the sea, however, heating of the lower air continues by day and night in winter months so that showers and squalls can occur at any time, and these may affect windward coastal areas. The average daily mean temperature at Kew with mP air and the difference between this and the average for all air masses (termed the *departure*) during the period 1931–45 are shown in Table 5.3. More extreme conditions occur with mA air, the temperature departures at Kew being approximately −4°C in summer and winter. The visibility in mA air is usually very good. The contribution of mP and mA air masses to the mean annual rainfall over a 5-year period at three stations in northern England is given in Table 5.4, although it should be noted that both air masses may also be involved in nonfrontal polar lows. Over much of southern England, and in areas to the lee of high ground, northerly and northwesterly airstreams usually give clear sunny weather with few showers. There is some indication of this in Table 5.4, for at Rotherham, in the lee of the Pennines, the percentage of the rainfall occurring with mP air is much lower than over Lancashire.

Maritime polar air which has followed a more southerly, cyclonic track over the Atlantic, approaching Britain from the southwest, or air which is moving northward ahead of a depression, is shown as mPw in Fig. 5.3 and Table 5.3. This air has surface properties intermediate with mT air and need not be discussed separately.

Table 5.3: Screen-level temperatures occurring with different air masses at Kew (°C) (*after Belasco, 1952*).

		mP (from Iceland)	mP (from mid-Atlantic)	mPw	mA	cP	mT	cT
Winter	T	4	6	8	0	−1	10	7
	D	−0.8	+1.3	+3.9	−4.4	−5.8	+5.8	+2.8
Summer	T	16	18	18	14	—	19	22
	D	−2.8	+0.6	−1.0	−3.3	—	+2.2	−4.4

T = Average mean daily temperature (to nearest whole degree Celsius).
D = Departure from the average for all air masses (1931–45).

Maritime tropical air commonly forms the warm sector of depressions moving from between west and south towards Britain, but Fig. 5.3 excludes cases of fronts and depressions (which have a frequency of about 10–12% throughout the year). Hence the characteristic weather conditions of mT air occur rather more often than the percentage frequency might suggest (in January 11% mT air and 4% Anticyclonic air originating from southwest of Britain). The weather is unseasonably mild and damp with mT air in winter. There is generally a complete cover of stratus or stratocumulus cloud and drizzle or light rain (formed by coalescence) may occur, especially over high ground where low cloud produces hill fog. The clearance of cloud on nights with light winds readily cools the moist air to its dew point, forming mist and fog. Table 5.4 shows that a large proportion of the annual rainfall is associated with warm-front and warm-sector situations and therefore is

Table 5.4: Percentage of the annual rainfall (1956–60) occurring with different synoptic situations (*after Shaw, 1962*).

Station	Warm Front	Warm Sector	Cold Front	Occlu- sion	Polar Low	mP	cP	Arctic	Thunder- storm
								Synoptic Categories	
Squires Gate (10 m or 33 ft)*	23	16	14	15	7	22	0.2	0.7	3
Greenfold Reservoir (305 m or 1000 ft)†	21	15	14	15	8	24	0.5	0.4	2
Rotherham (21 m or 70 ft)‡	26	9	11	20	14	15	1.5	1.1	3

* On the Lancashire coast (Blackpool).
† Reservoir in the Rossendale Uplands (Lancashire).
‡ In the Don valley, Yorkshire.

largely attributable to convergence and frontal uplift within mT air. In summer the cloud cover with this air mass keeps temperatures closer to average than in winter (Table 5.3); night temperatures tend to be high, but daytime maxima remain rather low.

True continental polar air only affects the British Isles between December and February and even then it is relatively infrequent. Mean daily temperatures are well below average and maxima rise to only a degree or so above freezing point. The air mass is basically very dry and stable but a track over the central part of the North Sea supplies sufficient heat and moisture to cause showers, often in the form of snow, over eastern England and Scotland. *In toto* this provides only a very small contribution to the annual precipitation, as Table 5.4 shows, and on the west coast the weather is generally clear. A transitional cP–cT type of air mass reaches Britain from southeast Europe in all seasons, though less frequently in summer. Such airstreams are dry and stable.

Continental tropical air occurs on average about one day per month in summer, which accounts for the rarity of summer heat-waves since these south or southeast winds bring hot, settled weather. The lower layers are stable and the air is commonly hazy, but the upper layers tend to be unstable and occasionally intense surface heating may trigger-off a thunderstorm. In winter such modified cT air sometimes reaches Britain from the Mediterranean bringing fine, hazy, mild weather.

4. Singularities and natural seasons. Most popular weather lore expresses the belief that each season has its own weather (for example, in England

"February fill-dyke", "April showers"), while ancient adages suggest that even the sequence of weather may be determined by the conditions established on a given date (for example, 40 days of wet or fine weather following St Swithin's day, July 15). Some of these ideas are quite fallacious, but others contain more than a grain of truth if properly interpreted.

The tendency for a certain type of weather to recur with reasonable regularity around the same date is termed a *singularity*. Many calendars of singularities have been compiled, particularly in Europe, but the early ones (which concentrated upon anomalies of temperature or rainfall) did not prove very reliable. More recently greater success has been achieved by studying singularities of circulation pattern and catalogues have been prepared for the British Isles by Lamb and for central Europe by Flohn and Hess. Lamb's results are based on calculations of the daily frequency of the airflow categories between 1898 and 1947, some examples of which are shown in Fig. 5.4. A noticeable feature is the infrequency of the Westerly type in spring, the driest season of the year in the British Isles and also in northern France, northern Germany and in the countries bordering the North Sea. The catalogue of Flohn and Hess is based on a classification of large-scale

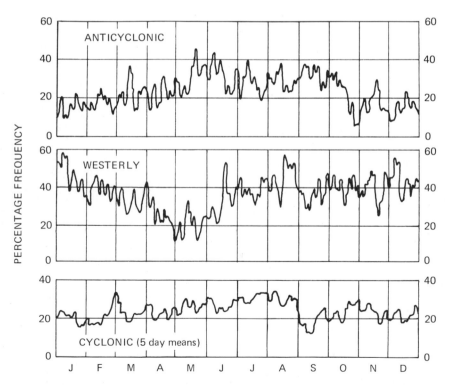

Fig. 5.4. The percentage frequency of anticyclonic, westerly, and cyclonic conditions over Britain, 1898–1947 (after Lamb, 1950).

patterns of airflow in the lower troposphere (*Grosswetterlage*) over central Europe originally proposed by F. Baur.

Some of the European singularities which recur most regularly are as follows:

(1) A sharp increase in the frequency of westerly and northwesterly type over Britain takes place about the middle of June. These invasions of maritime air also affect central Europe and this period is sometimes referred to as the beginning of the European "summer monsoon."

(2) About the second week in September Europe and Britain are affected by a spell of Anticyclonic weather. This may be interrupted by Atlantic depressions giving stormy weather over Britain in late September, though anticyclonic conditions again affect central Europe at the end of the month and Britain during early October.

(3) A marked period of wet weather often affects western Europe and also the western half of the Mediterranean at the end of October, whereas the weather in eastern Europe generally remains fine.

(4) Anticyclonic conditions return to Britain and affect much of Europe about mid-November, giving rise to fog and frost.

(5) In early December Atlantic depressions push eastward to give mild, wet weather over most of Europe.

In addition to these singularities, major seasonal trends are recognizable and for the British Isles Lamb identified five *natural seasons* on the basis of spells of a particular type lasting for 25 days or more during the period 1898–1947 (Fig. 5.5). In as far as it is possible to think in terms of a normal year, the seasons are:

(i) *Spring–early summer* (the beginning of April to mid-June). This is a period of variable weather conditions during which long spells are least likely. Northerly spells in the first half of May are the most significant feature, although there is a marked tendency for anticyclones to occur in late May–early June.

(ii) *High summer* (mid-June to early September). Long spells of various types may occur in different years. Westerly and Northwesterly types are the most common and they may be combined with either Cyclonic or Anticyclonic types. Persistent sequences of Cyclonic type occur more frequently than Anticyclonic ones.

(iii) *Autumn* (the second week in September to mid-November). Long spells are again present in most years, Anticyclonic ones mainly in the first half, Cyclonic and other stormy ones generally in October–November.

(iv) *Early winter* (from about the third week in November to mid-January).

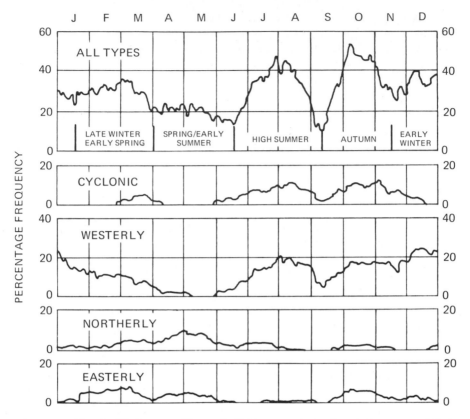

Fig. 5.5. The frequency of long spells (25 days or more) of a given air-flow type over Britain, 1898–1947. The diagram showing all long spells also indicates a division of the year into "natural seasons" (after Lamb, 1950).

Long spells are less frequent than in summer and autumn. They are usually of Westerly type giving mild, stormy weather.

(v) *Late winter and early spring* (from about the third week in January to the end of March). The long spells at this time of year can be of very different types, so that in some years it is mid-winter weather while in other years there is an early spring from about late February.

5. Synoptic anomalies. The mean climatic features of pressure and wind and the typical seasonal airflow regime provide only an incomplete picture of climatic conditions. Some patterns of circulation occur irregularly and yet, because of their tendency to persist for weeks or even months, form an essential element of the climate.

A major example is the occurrence of *blocking* patterns. It was noted in Chapter 3, Section E.2 that the zonal circulation in middle latitudes some times breaks down into a cellular pattern. This is commonly associated with a split of the jet stream into two branches over higher and lower middle latitudes and the formation of a cut-off low (see Chapter 4, Section F.4) south of a high-pressure cell. The latter is referred to as a *blocking anticyclone* since it prevents the normal eastward motion of depressions in the zonal flow. A major area of blocking is Scandinavia, particularly in spring. Depressions are forced to move northeastwards towards the Norwegian Sea or south-eastwards into southern Europe. This pattern, with easterly flow around the southern margins of the anticyclone, produces severe winter weather over much of northern Europe. In January–February 1947, for example, easterly flow across Britain as a result of blocking over Scandinavia led to extreme cold and frequent snowfall. Winds were almost continuously from the east between January 22 and February 22 and even daytime temperatures rose little above freezing point. Snow fell in some part of Britain every day from January 22 to March 17, 1947, and major snowstorms occurred as occluded Atlantic depressions moved slowly across the country. Other notably severe winter months—January 1881, February 1895, and January

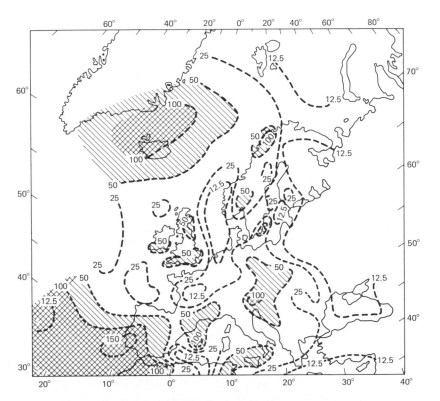

Fig. 5.6. The mean precipitation anomaly, as a percent of the average, during anticyclonic blocking in winter over Scandinavia. Areas above normal are cross-hatched, areas recording precipitation between 50 and 100% of normal have oblique hatching (after Rex, 1950).

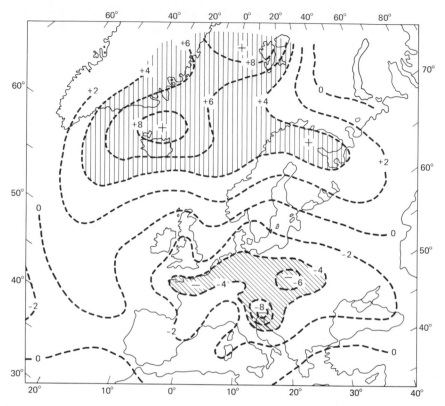

Fig. 5.7. The mean surface temperature anomaly (°C) during anticyclonic blocking in winter over Scandinavia. Areas more than 4°C above normal have vertical hatching, those more than 4°C below normal have oblique hatching (after Rex, 1950).

1940—were the result of similar pressure anomalies with pressure much above average to the north of the British Isles and below average to the south.

The average effects of a number of winter blocking situations over north-west Europe are shown in Figs. 5.6 and 5.7. Precipitation amounts are above normal mainly over Iceland and the western Mediterranean as depressions are steered around the blocking high following the path of the upper jet streams. Over most of Europe precipitation remains below average and this pattern is repeated with summer blocking. Winter temperatures are above average over the northeastern Atlantic and adjoining land areas, but below average over central and eastern Europe and the Mediterranean due to the northerly outbreaks of cP air (Fig. 5.7). The negative temperature anomalies associated with cool northerly airflow in summer cover most of Europe, and only northern Scandinavia and the northeastern Atlantic have above-average values.

Despite these generalizations the location of the block is of the utmost importance. For instance, in the summer of 1954 a blocking anticyclone across eastern Europe and Scandinavia allowed depressions to stagnate over the British Isles giving a dull, wet August, whereas in 1955 the blocking was located over the North Sea and a fine, warm summer resulted. Another less common location of blocking is Iceland. A notable example was the 1962–3 winter when persistent high pressure southeast of Iceland led to northerly and northeasterly airflow over Britain. Temperatures in central England at that time were the lowest since 1740, with a mean of 0°C for December 1962 to February 1963. Central Europe was affected by easterly airstreams and mean January temperatures there were 6°C below average.

6. Topographic effects. In various parts of Europe topography has a marked effect on the climate not only of the uplands themselves but also on that of adjacent areas. Apart from the more obvious effects upon temperatures, precipitation amounts, and winds, the major mountain masses affect the movement of frontal systems too. Surface friction over mountain barriers tends to steepen the slope of cold fronts and decrease the slope of warm fronts so that the latter are slowed down and the former accelerated. The cyclo-genetic effect of mountain barriers in producing lee depressions has already been discussed (see Chapter 4, Section F.1).

The Scandinavian mountains are one of the most significant climatic barriers in Europe as a result of their orientation with regard to westerly airflow. Maritime air masses are forced to rise over the highland zone giving annual precipitation totals of over 250 cm (100 in.) on the mountains of western Norway, whereas descent in their lee produces a sharp decrease in the amounts. The upper Gudbrandsdalen and Osterdalen in the lee of the Jotunheim and Dovre Mountains receive on average less than 50 cm (20 in.), and similar low values are recorded in the Jämtland province of central Sweden around Östersund.

The mountains can equally function in the opposite sense. For example, Arctic air from the Barents Sea may move southwards in winter over the Gulf of Bothnia, usually when there is a depression over northern Russia, giving very low temperatures in Sweden and Finland. Western Norway is rarely affected, since the cold wave is contained to the east of the mountains. In consequence there is a sharp climatic gradient across the Scandinavian Highlands in the winter months.

The Alps provide a quite different illustration of topographic effects. Together with the Pyrenees and the mountains of the Balkans, the Alps effectively separate the Mediterranean climatic region from that of Europe. The penetration of warm air masses north of these barriers is comparatively rare and short-lived. However, with certain pressure patterns, air from the Mediterranean and north Italy is forced to cross the Alps, losing its moisture through precipitation on the southern slopes. Dry adiabatic warming on the northern side of the mountains can readily raise temperatures by 5–6°C in the upper valleys of the Aar, Rhine, and Inn. At Innsbruck there are about 40 days of these föhn winds per year with a spring maximum, and these occurrences are often responsible for rapid melting of the snow, creating a risk of avalanches. When the airflow across the Alps has a northerly component föhns may occur in northern Italy, but their effects are generally less pronounced.

A more detailed examination of upland climates, with particular reference to Britain, will serve to illustrate some of the diverse effects of altitude. The mean annual rainfall on the west coasts near sealevel is about 114 cm (45 in.) but on the western mountains of Scotland, the Lake District, and Wales averages exceed 380 cm (150 in.) per year. The annual record is 653 cm (257 in.) in 1954 at Sprinkling Tarn, Cumberland, and 145 cm (57 in.) fell in a single month (October 1909) just east of the summit of Snowdon. The annual number of rain-days (days with at least 0.25 mm (0.01 in.) of precipitation) increases from about 165 days in southeastern England and the south coast to over 230 days in northwest Britain. There is little additional increase in the frequency of rainfall with height on the mountains of the Northwest, so that the mean rainfall per rain-day rises sharply from 0.5 cm (0.2 in.) near sea level in the West and Northwest to over 1.3 cm (0.5 in.) in the Western Highlands, the Lake District, and Snowdonia. This demonstrates that "orographic rainfall" here is primarily due to an intensification of the normal precipitation processes and is not a special type. It is more appropriate, therefore, to recognize an orographic component which increases the amounts of rain associated with frontal depressions and unstable airstreams (see Chapter 2, Section H.2). Even quite low hills such as the Chilterns and South Downs cause a rise in rainfall, receiving about 12–13 cm (5 in.) per year more than the surrounding lowlands. Indeed, detailed studies in Sweden show that wooded hills rising only 30–50 meters (100–150 ft) above the surrounding plains may cause precipitation amounts during cyclonic spells to be increased by 50–80% compared with the average falls over the lowland. However, in most countries, the rain-gauge networks are too coarse to detect such small-scale variations.

The sheltering effects of the uplands produce low annual totals on the lee side (with respect to the prevailing winds). Thus, the lower Dee valley in the lee of the mountains of north Wales receives less than 75 cm (30 in.) per year compared with over 250 cm (100 in.) on Snowdonia.

The complexity of the various factors affecting rainfall in Britain is shown by the fact that a close correlation exists between annual totals in northwest Scotland, the Lake District, and western Norway, which are directly affected by Atlantic depressions. At the same time there is an inverse relationship between annual amounts in the Western Highlands and lowland Aberdeenshire less than 240 km (150 miles) away. Annual precipitation in the latter area is more closely correlated with that in lowland eastern England.

Essentially, the British Isles comprise two major climatic units for rainfall: first, an Atlantic one with a winter season maximum, and, second, those central and eastern districts with continental affinities in the form of a weak summer maximum in most years. Other areas (eastern Ireland, eastern Scotland, northeast England, and most of midland England and the Welsh border counties) generally have a wet second half of the year.

The occurrence of snow is another measure of altitude effects. Near sea level there are on average approximately 5 days per year with snow falling in southwest England, 15 days in the southeast, and 35 days in northern Scotland. Between 60 and 300 meters the frequency increases by about 1 day per 15 meters of elevation and even more rapidly on higher ground. Approximate figures for northern Britain are 60 days at 600 meters and 90 days at 900 meters. The number of mornings with snow lying on the ground (more than half the ground covered) is closely related to mean temperature and hence altitude. Average figures range from about 5 days per year or less in much of southern England and Ireland, to between 30 and 50 days on the Pennines and over 100 days on the Grampian Mountains. In the last area (on the Cairngorms) and on Ben Nevis there are several semipermanent snowbeds at about 1160 meters and it is estimated that the theoretical climatic snowline—above which there is *net* accumulation of snow—is at 1620 meters (5300 ft) over Scotland.

The seasonal variability of lapse rates in mountain areas was mentioned in Chapter 1, Section H. There also exist marked geographical variations even within the British Isles. One measure of these variations is the length of the growing season. Meteorological data can be used to determine an index of growth opportunity by counting the number of days on which the mean daily temperature exceeds an arbitrary threshold value—commonly 6°C (43°F). Along the southwest coasts of England the growing season, as calculated on this basis, is nearly 365 days per year and in this area it decreases by about 9 days per 30 meters of elevation, but in northern England and Scotland the decrease is only about 5 days per 30 meters from between 250–270 days near sea level. In continental climates the altitudinal decrease may be even more gradual; in central Europe and New England, for example, it is about 2 days per 30 meters.

B. NORTH AMERICA

The North American continent spans nearly sixty degrees of latitude and, not surprisingly, exhibits a wide range of climatic conditions. Unlike Europe, the west coast is backed by the Pacific coast ranges rising to over 2,750 meters, which lie across the path of depressions in the mid-latitude westerlies and prevent the extension of maritime influences inland. In the interior of the continent there are no significant obstructions to air movement, and the absence of any east–west barrier allows air masses from the Arctic or the

Gulf of Mexico to sweep across the interior lowlands, causing wide extremes of weather and climate. Maritime influences in eastern North America are greatly limited by the fact that the prevailing winds are westerly, so that the temperature regime is continental. Nevertheless, the Gulf of Mexico is a major source of moisture supply for precipitation over the eastern half of the United States and, as a result, the precipitation regimes are different from those found in eastern Asia.

We will look first at the broad characteristics of the atmospheric circulation over the continent.

1. Pressure systems. The mean pressure pattern for the middle troposphere displays a prominent trough over eastern North America in both summer and winter (see Fig. 3.11). One theory is that this is a lee trough caused by the effect of the western mountain ranges on the upper westerlies, but at least in winter the strong baroclinic zone along the east coast of the

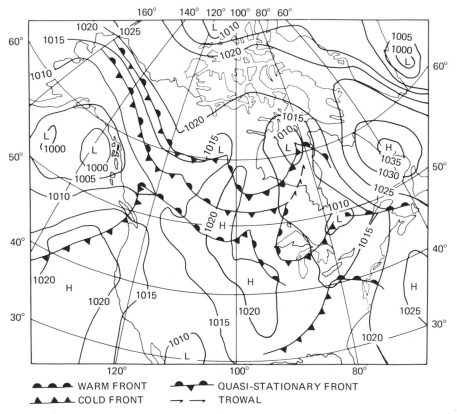

Fig. 5.8. A synoptic example of depressions associated with three-frontal zones on May 29, 1963 over North America (based on charts of the Edmonton Analysis Office and the Daily Weather Report).

continent is undoubtedly a major contributory factor. The implications of this mean wave pattern are that cyclones tend to move southeastward over the Midwest carrying continental polar air southward, while the cyclone paths are northeastward along the Atlantic coast.

In individual months there may of course be considerable deviations from this average pattern, with important consequences for the weather in different parts of the continent, and, in fact, this relationship provides the basis for the monthly forecasts of the United States Weather Bureau. For example, if the trough is more pronounced than usual temperature may be much below average in the central, southern, and eastern United States, whereas if the trough is weak the westerly flow is stronger with correspondingly less opportunity for cold outbreaks of polar air masses. Sometimes the trough is displaced to the western half of the continent causing a reversal of the usual weather pattern, since upper northwesterly airflow can bring cold, dry weather to the west while in the east there are very mild conditions associated with upper southwesterly flow. Precipitation amounts also depend on the depression tracks; if the upper trough is far to the west, depressions form ahead of it (see Chapter 4, Section G) over the south central United States and move northeastwards towards the lower St Lawrence, giving more precipitation than usual in these areas and less along the Atlantic coast.

The major features of the surface pressure map in January (see Fig. 3.16A) are the extension of the subtropical high over the southwestern United States (called the Great Basin high) and the separate polar anticyclone of the Mackenzie district of Canada. Mean pressure is low off both the east and west coasts of higher middle latitudes, where oceanic heat sources indirectly give rise to the (mean) Icelandic and Aleutian lows. It is interesting to note that, on average, in December, of any region in the northern hemisphere for any month of the year, the Great Basin region has the most frequent occurrence of highs, whereas the Gulf of Alaska has the maximum frequency of lows. The Pacific coast as a whole has its most frequent cyclonic activity in winter as does the Great Lakes area, whereas over the Great Plains the maximum is in spring and early summer. Remarkably, the Great Basin in June has the most frequent cyclogenesis of any part of the northern hemisphere in any month of the year. Heating over this area in summer helps to maintain a shallow, quasipermanent low-pressure cell, in marked contrast with the almost continuous subtropical high pressure belt in the middle troposphere (Fig. 3.11). Continental heating also indirectly assists in the splitting of the Icelandic low to create a secondary center over northeastern Canada. The west-coast summer circulation is dominated by the Pacific anticyclone, while the southeastern United States is affected by the Atlantic subtropical anticyclone cell.

Broadly, there are three prominent depression tracks across the continent in winter (see Fig. 4.14). One group moves from the west along a more or less zonal path about 45–50°N, whereas a second loops southward over the central United States and then turns northeastward towards New England and the Gulf of St Lawrence. Some of these depressions originate over the Pacific, cross the western ranges as an upper trough, and redevelop in the lee of the mountains. Alberta is a noted area for this process and also for primary cyclogenesis since the arctic frontal zone is over northwest Canada in winter. This frontal zone involves much-modified mA air from the Gulf of Alaska and cold dry cA (or cP) air. Depressions of the third group form along the main polar frontal zone, which in winter is off the east coast of the United States, and move northeastward towards Newfoundland. Sometimes this frontal zone is present over the continent about 35°N with mT air from the Gulf and cP air from the north or modified mP air from the Pacific. Polar front depressions forming over Colorado move northeastward towards the Great Lakes and others developing over Texas follow a more or less parallel path, further to the south and east, towards New England.

Between the Arctic and Polar Fronts a third frontal zone is distinguished by Canadian meteorologists. This *Maritime* (arctic) frontal zone is present when mA and mP (or mPc and mPw) air masses interact along their common boundary. The 3-front (or four air-mass) model allows a detailed analysis to be made of the baroclinic structure of depressions over the North American continent using synoptic weather maps and cross-sections of the atmosphere. Fig. 5.8 illustrates the three frontal zones and associated depressions on May 29, 1963. Along 95°W, from 60°N to 40°N, the following dew-point temperatures were reported in the four air masses: 17°F (-8°C), 33°F (1°C), 40°F (4°C), and 55°F (13°C).

In summer, the east-coast depressions are less frequent and the tracks across the continent are displaced northwards with the main ones moving over Hudson Bay and Labrador–Ungava, or along the line of the St Lawrence. These are associated mainly with a rather poorly defined Maritime frontal zone. The Arctic Front is usually located along the north coast of Alaska where there is a strong temperature gradient between the bare land and the cold Polar Sea and pack-ice. East from here the front is very variable in location from day to day and year to year. Broadly, it occurs most often in the vicinity of northern Keewatin and Hudson Strait, although one study of air mass temperatures and airstream confluence regions suggests that *an* arctic frontal zone occurs further south over Keewatin in July and that its mean position in this area is closely related to the boreal forest–tundra boundary. This relationship undoubtedly reflects the importance of arctic air-mass dominance for summer temperatures and consequently for tree-growth possibilities, but the precise nature of the interrelationships between atmospheric systems and vegetation boundaries requires more extensive investigation.

A number of circulation singularities have been recognized in North America, as in Europe (Chapter 5, Section A.4). Three which have received considerable attention in view of their prominence are: (i) the advent of spring in late March; (ii) the midsummer "high pressure jump" in late June; (iii) the Indian summer in late September (and late October).

The arrival of spring is marked by different climatic responses in different parts of the continent. For example, there is a sharp decrease in March to April precipitation in California, due to the extension of the Pacific high, whereas precipitation intensity increases in the Midwest (Fig. 5.9A) as a result of more frequent cyclogenesis in Alberta and Colorado and a northward extension of maritime tropical air over the Midwest from the Gulf of Mexico. These changes are part of a hemispheric readjustment of the circulation since at the beginning of April the Aleutian low-pressure cell, which from

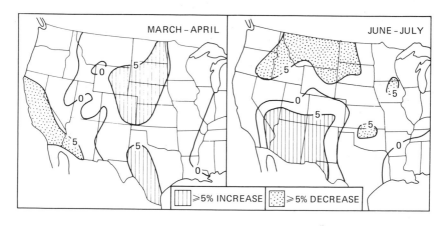

Fig. 5.9 The precipitation changes between March–April (left) and June–July (right), as a percentage of the mean annual total, for the central and western United States (after Bryson and Lahey, 1958).

September to March is located about 55°N, 165°W, splits into two with one center in the Gulf of Alaska and the other over northern Manchuria. This represents a decrease in the zonal index (Chapter 3, Section E.2).

In late June there is a rapid northward displacement of the subtropical high pressure cells in the northern hemisphere. In North America this pushes the depression tracks northward also with the result that precipitation decreases from June to July over the northern Great Plains, parts of Idaho and eastern Oregon (Fig. 5.9B). Conversely, the southwesterly anticyclonic flow which affects Arizona in June is replaced by air from the Gulf of Mexico flowing around the western side of the Bermuda high and this causes the onset of the summer rains (see Chapter 5, Section D.2). It has been suggested by Bryson and Lahey that these circulation changes at the end of June may be connected with the disappearance of snow cover from the arctic tundra. This leads to a sudden decrease of surface albedo from about 75% to 15% with consequent changes in the heat budget components and hence in the atmospheric circulation.

Frontal wave activity makes the first half of September a rainy period in the northern midwest states of Iowa, Minnesota, and Wisconsin, but after about the 20th of the month anticyclonic conditions return with warm air-flow from the dry southwest giving fine weather—the so-called Indian summer. Significantly, the hemispheric zonal index value rises in late September. This anticyclonic weather type has a second phase in the latter half of October, but at this time there are polar outbreaks. The weather is generally cold and dry, although if precipitation does occur there is a high probability of it being in the form of snow.

2. The temperate west coast and cordillera. The oceanic circulation of the North Pacific closely resembles that of the North Atlantic. The drift from the Kuro Shio current off Japan is propelled by the westerlies towards the western coast of North America and acts as a warm current between 40° and 60°N. Sea-surface temperatures are several degrees lower than in comparable latitudes off western Europe, however, due to the smaller volume of warm water involved. Also, in contrast to the Norwegian Sea, the shape of the Alaskan coastline prevents the extension of the drift to high latitudes.

The Pacific coast ranges greatly restrict the inland extent of oceanic influences, and hence there is no extensive maritime temperate climate such as we have described for western Europe. The major climatic features

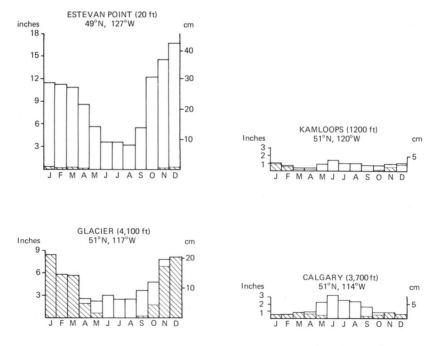

Fig. 5.10. Precipitation graphs for stations in western Canada. The shaded portions represent snowfall, expressed as water equivalent.

duplicate those of the coastal mountains of Norway and those of New Zealand and southern Chile in the belt of Southern Westerlies. Topographic factors make the weather and climate of such areas very variable over short distances both vertically and horizontally, and therefore only a few salient characteristics are selected for consideration.

There is a regular pattern of rainy windward and drier lee slopes across the successive northwest to southeast ranges with a more general decrease towards the interior. The Coast Range in British Columbia has mean annual totals of precipitation exceeding 250 cm (100 in.) with 500 cm (200 in.) in the wettest places compared with 125 cm (50 in.) or less on the summits of the Rockies, yet even on the leeward side of Vancouver Island the average figure at Victoria is only 70 cm (27 in.). Analogous to the Westerlies-oceanic regime of northwest Europe, there is a winter precipitation maximum along the littoral which also extends beyond the Cascades (in Washington) and the Coast Range (in British Columbia), but summers are drier due to the strong North Pacific anticyclone. The regime in the interior of British Columbia is transitional between that of the coastal region and the distinct summer maximum of central North America (Fig. 5.10), although at Kamloops in the Thompson valley (annual average 25 cm or 10 in.) there is a slight summer maximum associated with thunderstorm type of rainfall. In general, the sheltered interior valleys receive less than 50 cm (20 in.) per year and in the driest years certain localities have recorded only 15 cm (6 in.). Above 1000 meters much of the precipitation falls as snow (Fig. 5.10) and some of the greatest snow-depths in the world are reported from British Columbia, Washington, and Oregon. For example, between 1000–1500 cm (400–600 in.) falls on the Cascade Range at heights of about 1500 meters (5,000 ft) and even as far inland as the Selkirk Mountains the totals are considerable. The mean snowfall is 990 cm (390 in.) at Glacier, British Columbia (elevation 1200 meters) and this accounts for almost 70% of the annual precipitation. In Canada snow-depth is generally measured in inches, and ten inches of snow is taken to be equivalent to one inch of water. It is estimated that the climatic snowline rises from about 1600 meters (5250 ft) on the west side of Vancouver Island to 2900 meters (9500 ft) in the eastern Coast Range. Inland its elevation increases from 2300 meters (7550 ft) on the west slopes of the Columbia Mountains to 3100 meters (10,170 ft) on the east side of the Rockies. This trend reflects the precipitation pattern referred to above.

Finally, mention must be made of the large diurnal variations which affect the cordilleran valleys. Strong diurnal rhythms of temperature (especially in summer) and wind direction are a feature of mountain climates and their effect is superimposed upon the general climatic characteristics of the area. Cold air drainage produces many remarkably low minima in the mountain valleys and basins. At Princeton, British Columbia (elevation 695 meters) where the mean daily minimum in January is −14°C (7°F)

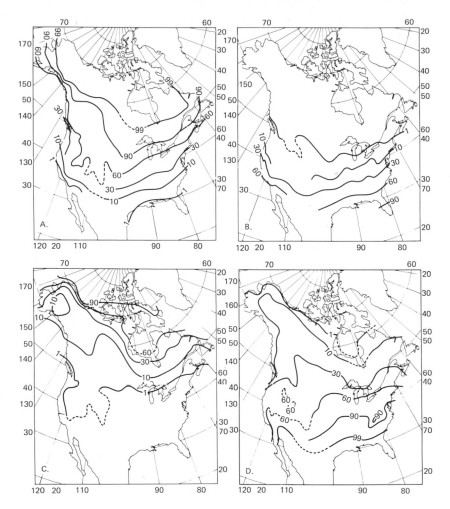

Fig. 5.11. The percentage frequency of hourly temperatures above or below certain limits for North America (after Rayner, 1961). A. January temperatures < 0°C. B. January temperatures > 10°C. C. July temperatures < 10°C. D. July temperatures > 21°C.

there is on record an absolute low of −45°C (−49°F), for example. This leads in some cases to reversal of the normal lapse rate. Golden in the Rocky Mountain Trench has a January mean of −12°C (11°F) whereas 460 meters (1500 ft) higher at Glacier (1248 meters) it is −10°C (14°F).

3. Interior and eastern North America. Central North America has the typical climate of a continental interior in middle latitudes with hot summers and cold winters (see Fig. 5.13), yet the weather in winter is subject to marked variability. This is determined by the steep temperature gradient between the Gulf of Mexico and the snow-covered northern plains; also by shifts of the upper wave patterns and jet stream. Cyclonic activity in winter is much more pronounced over central and eastern North America than in Asia, which is dominated by the Siberian anticyclone (see Fig. 4.14), and consequently there is no climatic type with a winter minimum of precipitation in eastern North America.

The general temperature conditions in winter and summer are illustrated in Fig. 5.11, showing the frequency with which hourly temperature readings exceed or fall below certain limits. The two chief features of all four maps are: (a) the dominance of the meridional temperature gradient, away from coasts, and (b) the continentality of the interior and east compared with the "maritimeness" of the west coast. On the July maps additional influences are evident and these are referred to below.

(a) Continental and oceanic influences. The Labrador coast is fringed by the waters of a cold current, analogous to the Oya Shio off eastern Asia, but in both cases the prevailing westerlies greatly limit their climatic significance. The Labrador Current maintains drift ice off Labrador and Newfoundland until June and gives very low summer temperatures along the

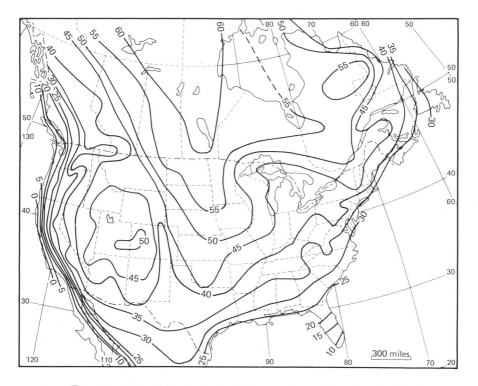

Fig. 5.12. Continentality in North America according to Conrad's index (modified after Trewartha, 1961).

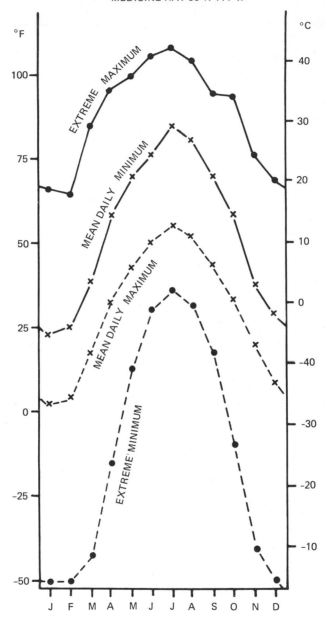

Fig. 5.13. Mean and extreme temperatures at Medicine Hat, Alberta.

Labrador coast (Fig. 5.11C). The lower incidence of freezing temperatures in this area in January is related to the movement of some depressions into the Davis Strait, carrying Atlantic air northwards. A major role of the Labrador current is in the formation of fog. Advection fogs are very frequent between May and August off Newfoundland, where the Gulf Stream and Labrador Current meet. Warm, moist southerly airstreams are cooled rapidly over the cold waters of the Labrador Current and with steady, light winds such fogs may persist for several days, creating hazardous conditions for shipping. Southward-facing coasts are particularly affected and at Cape Race (Newfoundland), for example, there are on average 158 days per year with fog (visibility less than 1 km) at some time of day. The summer concentration is shown by the figures from Cape Race during individual months: May—18 (days), June—18, July—24, August—21, and September—15.

Oceanic influence along the Atlantic coasts of the United States is very limited, and although there is some moderating effect on minimum temperatures at coastal stations this is scarcely evident on generalized maps such as Fig. 5.11. More significant climatic effects are in fact found in the neighbor-

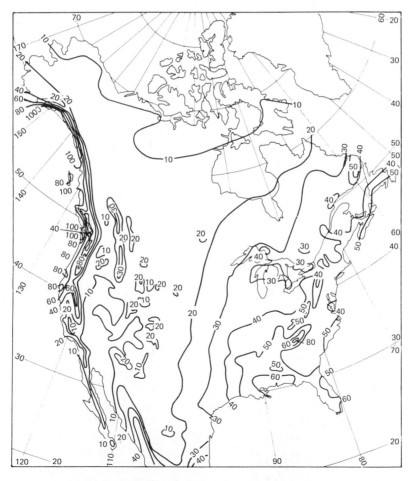

Fig. 5.14. Average annual precipitation (in inches) over North America (after Brooks and Conner, Kendrew, and Thomas.)

hood of Hudson Bay and the Great Lakes. Hudson Bay remains very cool in summer with water temperatures of about 7°–9°C and this depresses temperatures along its shore, especially in the east (Fig. 5.11C and D). Mean July temperatures are 12°C (54°F) at Churchill (59°N) and 8°C (47°F) at Port Harrison (58°N), on the west and east shores respectively, compared for instance with 13°C (56°F) at Aklavik (68°N) on the Mackenzie delta. The influence of Hudson Bay is even more striking in early winter when the land is snow-covered. Westerly airstreams crossing the open water are warmed on average by 11°C (20°F) in November, and moisture added to the air leads to considerable snowfall in western Ungava (see the graph for Port Harrison, Fig. 5.15). By the beginning of January the Bay is frozen over almost entirely and no effects are evident. The Great Lakes influence their surroundings in much the same way. Heavy winter snowfalls are a notable feature of the southern and eastern shores of the Great Lakes. In addition to contributing moisture to northwesterly streams of cold cA and cP air, the heat source of the open water in early winter produces a low-pressure trough which increases the snowfall as a result of convergence. Yet a further factor is frictional convergence and orographic uplift at the shore-line. Mean annual snowfall exceeds 250 cm (100 in.) along much of the eastern shore of Lake Huron and Georgian Bay, the southeastern shore of Lake Ontario, the northeastern shore of Lake Superior, and its southern shore east of about 90° 30′W. Extremes include 114 cm (45 in.) in one day at Watertown, New York, and 894 cm (352 in.) during the 1946–7 winter season at nearby Bennetts Bridge, both of which are close to the eastern end of Lake Ontario. Transport in cities in these snow belts is quite frequently disrupted during winter snowstorms. The Great Lakes also provide an important tempering influence during winter months in raising average daily minimum temperatures at lakeshore stations by some 2° to 4°C above those at inland locations.

An indication of the seasonal range of temperature is provided by Fig. 5.12, showing continentality (k) based on Conrad's formula:

$$k = \frac{1.7A}{\sin(\phi + 10)} - 14,$$

where A is the average annual temperature range in °C and ϕ is the latitude angle. The results in middle and high latitudes are similar to those obtained by Gorczynski's method (see Chapter 6, Section A); with either of these empirical expressions it is only the relative magnitude of k that is of interest. The highest values form a tongue along the 100°W meridian with subsidiary areas on the 'Lake Plateau' of central Labrador-Ungava and on the high plateaus of Colorado and Utah. The maritimeness of the Pacific coast, though of very limited inland extent, is pronounced, whereas on the east coast there is relatively high continentality. The map also illustrates the ameliorating effect of the Great Lakes.

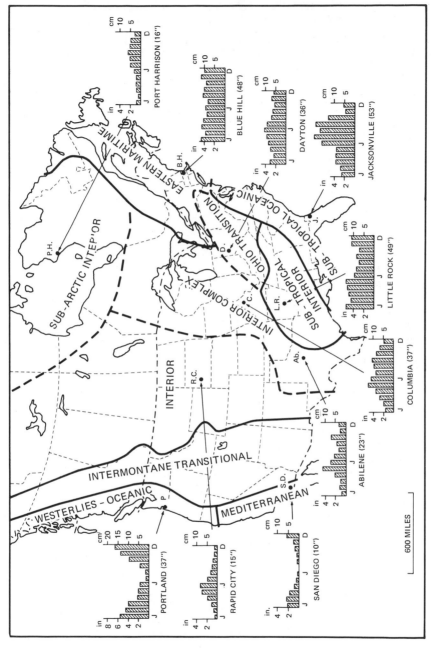

Fig. 5.15. Rainfall regimes in North America. The insert histograms show the mean monthly precipitation; January, June, and December are indicated (partly after Trewartha, 1961).

(b) Warm and cold spells. Two types of synoptic condition are of particular significance for temperatures in the interior of North America. One is the cold wave caused by a northerly outbreak of cP air, which in winter regularly penetrates deep into the central and eastern United States and occasionally affects even the Gulf coast (Fig. 5.11A), injuring frost-sensitive crops. Cold waves are arbitrarily defined as a temperature drop of at least 11°C (20°F) in 24 hours over most of the United States, and at least 9°C (16°F) in California, Florida, and the Gulf Coast, to below a specified minimum depending on location and season. The winter criterion decreases from 0°C in California, Florida, and the Gulf Coast to −18°C (0°F) over the northern Great Plains and the northeastern States. The cold spells commonly occur with the build-up of a north-south anticyclone in the rear of a cold front. The polar air gives clear, dry weather with strong, cold winds, although if the winds follow snowfall, fine, powdery snow may be whipped up by the wind, creating blizzard conditions. These are quite common over the northern plains.

Another type of temperature fluctuation is associated with the *Chinook* winds in the lee of the Rockies (see Chapter 3, Section B.2). The Chinook is particularly warm and dry as air of Pacific origin, after losing its moisture over the mountains, descends the eastern slopes and warms at the dry adiabatic lapse rate. The onset of the Chinook produces temperatures well above the seasonal normals so that snow is often thawed rapidly, and in fact the Indian word Chinook means snow-eater. Temperature rises of as much as 22°C (40°F) have been observed in five minutes and the occurrence of such warm spells is reflected by the high extreme maxima in winter months at Medicine Hat (Fig. 5.13). In Canada the Chinook effect may be observed a considerable distance from the Rockies into southwest Saskatchewan, but in Colorado its influence is rarely felt more than about 50 km from the foothills.

No adequate definition of a Chinook has yet been established, but using an arbitrary criterion of winter days with a maximum temperature of at least 4.4°C (40°F), R. W. Longley has shown that in the Lethbridge area Chinooks occur on 40% of days during December–February.

However, since the phenomenon is the result of a particular type of airflow, it is evident that some wind characteristic should be included in future definitions.

Chinook conditions commonly develop in a Pacific airstream which is replacing a winter high-pressure cell over the high western plains. Sometimes the cold, stagnant cP air of the anticyclone is not dislodged by the descending Chinook and a marked inversion is formed, but on other occasions the boundary between the two air masses may reach ground level locally and, for example, the western suburbs of Calgary may record temperatures above 0°C while those to the east of the city remain below −15°C.

(c) Precipitation and the moisture balance. Longitudinal influences are apparent in the distribution of annual precipitation, although this is in large measure a reflection of the topography. The 20 in. (51 cm) annual isohyet in the United States approximately follows the 100°W meridian (Fig. 5.14), and westwards to the Rockies is an extensive dry belt in the rain shadow of the western mountain ranges. In the southeast totals exceed 50 inches (127 cm), and 40 inches or more is received along the Atlantic coast as far as New Brunswick and Newfoundland.

The major sources of moisture for precipitation over North America are the Pacific Ocean and the Gulf of Mexico. The former need not concern us here since comparatively little of the precipitation falling over the interior appears to be derived from that source. The Gulf source is extremely important in providing moisture for precipitation over central and eastern North America, but the predominance of southwesterly airflow means that little precipitation falls over the western Great Plains (Fig. 5.14). Over the southeastern United States there is considerable evapotranspiration and this helps to maintain moderate annual totals northwards and eastwards from the Gulf by providing additional water vapor for the atmosphere. Along the east coast the Atlantic Ocean is an additional significant source of moisture for winter precipitation.

There are at least eight major types of seasonal precipitation regime in North America (Fig. 5.15); the winter maximum of the west coast and the transition type of the intermontane region in middle latitudes have already been mentioned and the subtropical types are discussed in the next section. Four primarily mid-latitude regimes are distinguished east of the Rocky Mountains:

(i) A warm season maximum is found over much of the continental interior (for example, Rapid City). In an extensive belt from New Mexico to the Prairie Provinces more than 40% of the annual precipitation falls in summer. In New Mexico the rain occurs mainly with late summer thunderstorms, but May–June is the wettest time over the central and northern Great Plains due to more frequent cyclonic activity. Winters are quite dry over the Plains, but the mechanism of the occasional heavy snowfalls is of interest. They occur over the northwestern Plains during easterly upslope flow, usually in a ridge of high pressure. Further north in Canada the annual maximum is commonly in late summer or autumn when depression tracks are in higher middle latitudes. There is a local maximum in autumn on the eastern shores of Hudson Bay (for example, Port Harrison) due to the effect of open water.

(ii) Eastward and southward of the first zone there is a double maximum in May and September. In the upper Mississippi region (for example, Columbia) there is a secondary minimum, paradoxically in July–August when the air is especially warm and moist, and a similar profile occurs in northern Texas (for example, Abilene). An upper-level ridge of high pressure over the Mississippi valley seems to be responsible for reduced thunderstorm rainfall in mid-summer, and a tongue of subsiding dry air extends southwards from this ridge towards Texas. In September, renewed cyclonic activity associated with the seasonal southward shift of the polar front, at a time when mT air from the Gulf is still warm and moist, causes a resumption of rainfall. Subsequently, however, drier westerly airstreams affect the continental interior as the general airflow becomes more zonal.

The diurnal occurrence of precipitation in the central United States is rather unusual for a continental interior. Sixty percent or more of the summer precipitation in Kansas, Nebraska, and Iowa occurs in the form of thundery downpours between 1800 and 0600 hours. It has been suggested that the nocturnal thunderstorms and rainfall are caused by a slow, large-scale circulation over the plains east of the Rocky Mountains, with a tendency for low-level divergence and subsidence by day and for convergence and rising air by night. In this view the mountains are an integral element of the system, but another explanation invokes north-south pulses of air to account for the necessary diurnal variations in the vertical motion over the plains. No generally accepted hypothesis has yet been put forward.

(iii) East of the upper Mississippi, in the Ohio valley and south of the lower Lakes, there is a transitional regime between that of the interior and the east coast type. Precipitation is reasonably abundant in all seasons but the summer maximum is still in evidence (for example, Dayton).

(iv) In the eastern North America (New England, the Maritimes, Quebec, and southeast Ontario) precipitation is fairly evenly distributed throughout the year (for example, Blue Hill). In Nova Scotia and locally around Georgian Bay there is a winter maximum, due in the latter case to the influence of open water. In the Maritimes it is related to winter (and also autumn) storm tracks. It is worth comparing the eastern regime with the summer maximum which is found over eastern Asia. There the Siberian anticyclone excludes cyclonic precipitation in winter and monsoonal influences are felt in the summer months.

The seasonal distribution of precipitation is of vital interest for agricultural purposes. Rain falling in summer, for instance, when evaporation losses are high is less effective than an equal amount in the cool season. Fig. 5.16 illustrates the effect of different regimes in terms of the moisture balance,

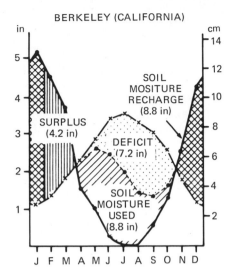

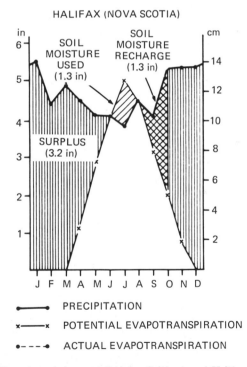

Fig. 5.16. The moisture balances at Berkeley, California and Halifax, Nova Scotia (after Thornthwaite and Mather, 1955, and Putnam et al., 1957).

calculated according to Thornthwaite's method. At Halifax (Nova Scotia) there is sufficient moisture stored in the soil to maintain evaporation at its maximum rate (or actual evaporation = potential evaporation), whereas at Berkeley (California) there is a computed moisture deficit of nearly 2 in. in August. This is a guide to the amount of irrigation water which may be required by crops, although in dry regimes the Thornthwaite method generally underestimates the real moisture deficit.

The mean monthly potential evaporation (PE) is calculated in the method developed by C. W. Thornthwaite from tables based on a complex equation relating PE to air temperature. This only gives a general guide to the true PE in view of the different factors which affect evaporation (see Chapter 2, Section A), but the results are reasonably satisfactory in temperate latitudes. By determining the annual moisture surplus and the annual moisture deficit from graphs such as those in Fig. 5.16, or from a monthly balance sheet, Thornthwaite obtained an index of aridity and humidity. The humidity index is $100 \times$ water surplus/PE and the aridity index is $100 \times$ water deficit/PE; there is generally a surplus in one season and a deficit in another. The humidity index and the aridity index may then be combined into a single moisture index (Im):

$$Im = \frac{100 \times \text{water surplus} - 60 \times \text{water deficit}}{PE}.$$

Here the deficit is given less weight as the surplus is held in the soil, whereas any deficit means that the actual evaporation rate falls below the potential value. The moisture index is used to define the following climatic types:

Im	Climate	Symbol
>100	Perhumid	A
20 to 100	Humid	B (with 4 subdivisions)
0 to 20	Moist subhumid	C_2
−20 to 0	Dry subhumid	C_1
−40 to −20	Semi-arid	D
<−40	Arid	E

Fig. 5.17 illustrates the distribution of these moisture regions over the United States. The zero line separating the moist climates of the east from the dry climates of the west (apart from the west coast) follows the 96th meridian almost exactly. The major humid areas are along the Appalachians, in the northeast, and along the Pacific coast, while the most extensive arid area is the southwest. Some aspects of the precipitation climatology of this arid area are examined in Chapter 5, Section D.2.

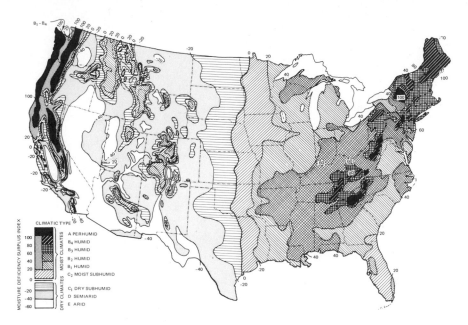

Fig. 5.17. Moisture regions of the United States. The moisture index, Im = [(100 × surplus) − (60 × deficit)]/PE (from Thornthwaite and Mather, 1955).

C. THE POLAR MARGINS

The longitudinal differences in mid-latitude climates persist into the polar margins, giving rise to maritime and continental subtypes, modified by the extreme radiation conditions in winter and summer. For example, insolation receipts in summer along the arctic coast of Siberia compare favorably, by virtue of the long daylight, with those in lower middle latitudes. The maritime type is found in coastal Alaska, Iceland, northern Norway, and adjoining parts of the U.S.S.R. Winters are cold and stormy, with very short days. Summers are cloudy but mild with mean temperatures about 10°C. For example, Vardø in north Norway (70°N, 31°E) has monthly mean temperatures of −6°C (21°F) in January and 9°C (48°F) in July, while Anchorage, Alaska (61°N, 150°W) records −11°C (12°F) and 14°C (57°F), respectively. Annual precipitation is generally between 60 and 125 cm (25 and 50 in.), with a cool season maximum and about six months of snow cover.

The weather is mainly controlled by depressions which are weakly developed in summer. In winter the Alaskan area is north of the main depression tracks and occluded fronts and upper troughs (trowals) are prominent, whereas northern Norway is affected by frontal depressions moving in to the Barents Sea. Iceland is similar to Alaska, though depressions often move slowly over the area and occlude, whereas others moving northeastwards along the Denmark Strait bring mild, rainy weather.

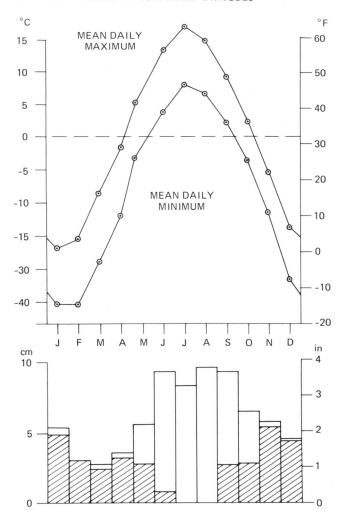

Fig. 5.18. Selected climatological data for McGill Sub-Arctic Research Laboratory, Knob Lake, 1955–62 (data from J. B. Shaw and D. G. Tout).

The interior, cold-continental climates have much more severe winters although precipitation amounts are less. At Yellowknife (62°N, 114°W), for instance, the mean January temperature is only −28°C (−18°F). In these regions *permafrost* (permanently frozen ground) is widespread and often of great depth. In summer only the top few feet of ground thaw and as the water cannot readily drain away this active layer often remains waterlogged. Although frost may occur in any month, the long summer days usually give three months with mean temperatures above 10°C, and at many stations extreme maxima are 32°C (90°F) or more (Fig. 5.11D). The Barren Grounds of Keewatin, however, are much cooler in summer due to the extensive areas of lake and muskeg, and only July has a mean daily temperature of 10°C. Labrador-Ungava to the east is rather similar, with very high cloud amounts and maximum precipitation in June–September (Fig. 5.18). In winter conditions fluctuate between periods of very cold, dry, high-pressure weather and spells of dull, bleak, snowy weather as depressions move eastwards or occasionally northwards over the area. In spite of the very low mean temperatures in winter, there have been occasions when maxima have exceeded 4.5°C (40°F) during incursions of maritime Atlantic air. Such variability is not found in eastern Siberia which is intensely continental, apart from the Kamchatka peninsula, with the northern hemisphere's *cold pole* located in the remote northeast (see Fig. 1.14). Verkhoyansk and Oimyakon have a January mean of −50°C and both have recorded an absolute minimum of −67.7°C.

D. THE SUBTROPICAL MARGINS

1. The Mediterranean. The characteristic west coast climate of the subtropics is the Mediterranean type with hot, dry summers and mild, relatively wet winters. It is interposed between the temperate maritime type and the arid subtropical desert climate, but the Mediterranean regime is transitional in a special way for it is controlled by the westerlies in winter and by the subtropical anticyclone in summer. The type region is peculiarly distinctive, extending more than 3000 km into the Eurasian continent. Additionally, the configuration of seas and peninsulas produces great regional variety of weather and climate. The Californian region with similar conditions

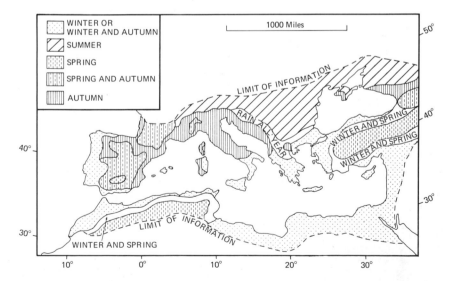

Fig. 5.19. Rainfall regimes in the Mediterranean area (after Huttary, 1950).

(Fig. 5.15) is of very limited extent, and attention is therefore concentrated on the Mediterranean basin itself.

In winter the Mediterranean is frequently affected by depressions (see Fig. 4.14) associated with a branch of the westerly jet stream located about 35°N. This develops during low index phases when the westerlies over the eastern Atlantic are distorted by a blocking anticyclone about 20°W, leading to a north or northwest airflow over the British Isles and France. Atlantic depressions may enter the Western Basin or, more frequently, cyclogenesis occurs in the Gulf of Genoa-Tyrrhenian Sea area in the lee of the Alps (see Chapter 4, Section F.1). Nearly 70% of all Mediterranean depressions are of this *Genoa type*. Shallow mP air can enter the area via the Rhone valley and Carcassonne gap, whilst the main airflow is checked by the Alps and Pyrenees, and this cold air forms a cold front in the incipient lee depression. The warm surface water (about 2°C warmer than the mean air temperature in January) provides an additional factor favoring cyclone development in cold north-westerly airstreams. This factor also causes rainfall of the shower and thunder-storm variety in the rear of depressions to be the dominant type affecting the Mediterranean region between about 5° and 25°E. As the depressions move eastward the mP (or mA) air behind the cold front is modified so that its characteristics approach those of air designated as *Mediterranean* (see Table 4.1). Around Cyprus, however, the depressions may be regenerated by fresh cP air from southern Russia, the Balkans, and Anatolia. Other lee depressions form south of the Atlas Mountains during northwesterly flow, and these are important sources of rainfall for north Africa, in spring espe-cially. With high index zonal circulation over the Atlantic and Europe depressions pass over northwestern Europe and weather in the Mediterranean is generally settled and fine. Between October and April, anticyclones are the dominant circulation type for at least 25% of the time over the whole Mediterranean area and in the western basin for 48% of the time. Conse-quently, although the winter half-year is the rainy period, there are rather few rain days. On average, rain falls on only 6 days per month during winter in northern Libya and southeast Spain, yet there are 12 rain days per month in western Italy, the western Balkan peninsula, and the Cyprus area. The higher frequencies (and totals) are related to the areas of cyclogenesis and to the windward side of peninsulas.

Regional winds are also related to the meteorological and topographic factors. The familiar cold, northerly winds of the Gulf of Lions (the *mistral*) and the northern Adriatic (the *bora*) occur when a depression is developing in the Gulf of Genoa east of a high-pressure ridge from the Azores anticyclone. Katabatic and funneling effects strengthen the flow in the Rhone valley and similar localities so that violent winds are sometimes recorded. The mistral may last for several days until the outbreak of polar or continental air ceases. The frequency of these winds depends on their definition. The average fre-quency of strong mistral in the south of France is shown in Table 5.5 (based on occurrence at one or more stations from Perpignan to the Rhone in 1924–7).

Table 5.5: Number of days with strong mistral in the south of France
(*after "Weather in the Mediterranean," H.M.S.O. 1962*).

Speed	J	F	M	A	M	J	J	A	S	O	N	D	Year
>11 m/sec (21 kt)	10	9	13	11	8	9	9	7	5	5	7	10	103
>17 m/sec (33 kt)	4	4	6	5	3	2	0.6	1	0.6	0	0	4	30

In summer the Mediterranean basin is dominated by the expanded Azores anticyclone in the west, while to the south the mean pressure field shows a low-pressure trough extending across Sahara from southern Asia. The winds are predominantly from a northerly direction (for example, the *Etesians* of the Aegean) and represent an eastward continuation of the northeasterly trades. Locally, sea breezes reinforce these winds, but on the Levant coast they cause surface southwesterlies. Depressions are by no means absent in the summer months, but they are usually weak since the anticyclonic character of the large-scale circulation encourages subsidence, and air-mass contrasts are much reduced compared with winter (see Table 4.2). Thermal lows form from time to time over Iberia and Anatolia, though thundery outbreaks are infrequent due to the low relative humidity.

The most important regional winds in summer are of continental tropical origin. There is a variety of local names for these hot, dry, and dusty airstreams—*Scirocco* (Algeria and the Levant), *Leveche* (southeast Spain), and *Khamsin* (Egypt)—which move northwards ahead of eastward-moving depressions.

Many stations in the Mediterranean receive only a few tenths of an inch of rainfall in at least one summer month, yet it is important to realize that the seasonal distribution does not conform to the pattern of simple winter maximum over the whole of the Mediterranean basin. Fig. 5.19 shows that this is found in the eastern and central Mediterranean, whereas Spain, southern France, northern Italy, and the northern Balkans have more complicated profiles with a maximum in autumn or peaks in both spring and autumn. This double maximum can be interpreted generally as a transition between the continental interior type with summer maximum and the Mediterranean type with winter maximum. A similar transition region occurs in the southwestern United States (see Fig. 5.15), but local topography in this intermontane zone introduces further irregularities into the regimes.

2. The semi-arid southwestern United States. Both the mechanisms and patterns of the climates of areas dominated by the subtropical high-pressure cells are rather obscure at present. On the one hand the inhospitable nature of these arid regions inhibits data collection, and on the other the proper interpretation of the irregular meteorological events requires a close network of stations maintaining continuous records over long periods. This difficulty is especially apparent in the interpretation of desert precipitation data, in that much of the rain falls in very local storms irregularly scattered both in space and time. It is convenient to treat aspects of this climatic type here, in that most of the reliable data relate to the less arid regions marginal to the subtropical cell centers, and in particular to the southwestern United States.

A series of observations at Tucson, Arizona, 730 meters (2400 ft) between 1895 and 1957 showed a mean annual precipitation of 27.7 cm (10.91 in.) falling on an average of about 45 days per year, with extreme mean annual figures of 61.4 cm (24.17 in.) and 14.5 cm (5.72 in.). Two moister periods of late November to March (receiving 30% of the mean annual precipitation) and late June to September (50%) are separated by more arid seasons from April to June (8%) and October to November (12%). The winter rains are generally prolonged and of low intensity (more than half the falls have an intensity of less than 0.5 cm (0.2 in.) per hour, falling from altostratus clouds at about 2500 meters (8000 ft) associated with the cold fronts of depressions which are forced to take southerly routes by strong blocking to the north. These southerly tracks occur during phases of equatorial displacement of the Pacific subtropical high-pressure cell. The reestablishment of the cell in spring before the main period of intense surface heating and convectional showers is associated with the most persistent drought periods, particularly during April to June. Dry westerly to southwesterly flow from the eastern edge of the Pacific subtropical anticyclone is responsible for the low rainfall at this season. During one 29-year period in Tucson there were 8 spells of more than 100 consecutive days of complete drought and 24 periods of more than 70 days.

The period of summer precipitation is quite sharply defined, beginning in the last week in June and lasting until the middle of September. At this time precipitation mainly occurs from convective cells initiated by surface heating, convergence, or, less commonly, orographic lifting. These summer convective storms form in clusters many tens of miles across, the individual storm cells covering together less than 3% of the surface area at any one time, and persisting for less than an hour on average. The storm clusters move across the country in the direction of the upper-air motion, and often seem to be controlled in movement by the existence of low-level jet streams at elevations of about 2500 meters. The airflow associated with these storms is generally southeasterly along the southern and western margins of the Atlantic (or Bermudan) subtropical high, so that in contrast to the winter months the moisture is derived mainly from the Gulf of Mexico. This circulation often sets in abruptly around July 1 and it is therefore recognized as a singularity (see Chapter 5, Section A.4. and Fig. 5.9B).

Precipitation from these cells is extremely local, and commonly concentrated in the midafternoon and evening. Intensities are much higher than in winter, half the summer rain falling at more than 0.4 inches per hour. During the 29-year period about one quarter of the mean annual precipitation fell in storms giving 2.5 cm (1 in.) or more per day, and 1.9 cm (0.75 in.) in 15 minutes were recorded. These intensities are much less than those associated with rainstorms in the humid tropics, but the sparsity of vegetation in the drier regions allows the rain to produce considerable surface erosion. Thus the highest measurements of surface erosion in the United States are from areas having 30–40 cm (12–15 in.) of rain per year.

3. The interior and east coast of the United States. The climate of the subtropical southeastern part of the United States has no exact counterpart in Asia, which is affected by the summer and winter monsoon systems. These are discussed in the next chapter and only the distinctive features of the North American subtropics are examined here. Seasonal wind changes are experienced in Florida, which is within the westerlies in winter and lies on the northern margin of the tropical easterlies in summer, but this is not comparable with the regime in southern and southeastern Asia. Nevertheless, the summer season rainfall maximum (Fig. 5.15 for Jacksonville) is a result of this changeover. In June the upper flow over the Florida peninsula changes from northwesterly to southerly as a trough moves westward and becomes established in the Gulf of Mexico. This deep, moist southerly airflow provides appropriate conditions for convection, and indeed Florida probably ranks as the state with the highest annual number of days with thunderstorms–90 or more, on average, in the vicinity of Tampa. These often occur in late afternoon although two factors apart from diurnal heating are thought to be important. One is the effect of sea breezes converging from both sides of the peninsula, and the other is the northward penetration of disturbances in the easterlies (see Chapter 6). The latter may of course affect the area at any time of day. The westerlies resume control in September–October, although Florida remains under the easterlies during September when Caribbean hurricanes are most frequent and consequently the rainy season is prolonged.

The region of the Mississippi lowlands and the southern Appalachians to the west and north is not simply transitional to the interior type in terms at least of rainfall regime (Fig. 5.15). The profile shows a winter–spring maximum and a secondary summer maximum. The cool season peak is related to westerly depressions moving northeastward from the Gulf coast area, and it is significant that the wettest month is commonly March when the mean jet stream is farthest south. The summer rains are associated with convection in humid air from the Gulf, though this convection becomes less effective inland as a result of the subsidence created by the anticyclonic circulation in the middle troposphere referred to previously (see Chapter 5, Section B.3.c).

6

Tropical Weather and Climate

Fifty percent of the surface of the globe lies between latitudes 30°N and 30°S and over a third of the world's population inhabits tropical lands. Tropical climates are therefore of especial geographical interest.

The latitudinal limits of tropical climates vary greatly with longitude and season, and tropical weather conditions may reach well beyond the Tropics of Cancer and Capricorn. For example, the summer monsoon extends to 30°N in southern Asia, but only 20°N in West Africa, while in late summer and autumn tropical hurricanes may affect extratropical areas of eastern Asia and eastern North America. Not only do the tropical margins extend seasonally polewards, but in the zone between the major subtropical high-pressure cells there is frequent interaction between temperate and tropical disturbances. Hence the tropical atmosphere is far from being a discrete entity and any meteorological or climatological boundaries must be arbitrary. There are, nevertheless, a number of distinctive features of tropical weather and these are discussed below.

A. THE ASSUMED SIMPLICITY OF TROPICAL WEATHER

The study of tropical weather has passed broadly through three stages. Firstly, for a long period which only ended some years before the Second World War, tropical weather conditions, patterns, and mechanisms were assumed to be much more simple and obvious than those in higher latitudes. This belief was partly due to the paucity of prewar meteorological records, particularly over the vast tropical oceans, and partly due to certain theoretical and practical considerations. One reason was that temperature and hence air-mass contrasts seemed small compared with middle latitudes. Air masses were nevertheless identified on the basis of their moisture content, temperature, and stability, although frontal activity was believed to be weak and

218

weather systems correspondingly less evident. Obvious exceptions were tropical cyclones which were thought to result from special conditions of thermal convection. Another reason was the large extent of ocean surface which, it was assumed, would simplify the patterns of weather and climate.

Thus the following simple picture of trade-wind weather evolved. Maritime tropical air masses, originating by subsidence in the subtropical high-pressure cells over the eastern halves of the oceans (see Fig. 3.19), move steadily westwards and equatorwards with quite constant speed and direction. Beneath the temperature inversion, formed between about 600 and 800 meters by subsidence, the air is moist with a layer of broken cumulus cloud. Conditions are invariably warm and dry, although in certain coastal regions (for example, southeast Madagascar) the onshore winds can produce periods of thick drizzle. Over the equatorial oceans the surface wind is light and variable (the doldrums) and the air is universally hot, humid, and sultry (see Chapter 3, Section D.1).

A further element of simplicity was thought to be provided by the insolation regime. The persistently high altitude of the sun in low latitudes and the equality of length between the days and nights means that seasonal variations of insolation are minimized. Hence the annual rhythm was thought to produce simple rainfall regimes with a single maximum following the summer solstice at the tropics, and a double peak at the equator in response to the passage of the overhead sun at the equinoxes. Diurnal patterns of land and sea breezes giving an afternoon build-up of convective activity and thundery showers were regarded as characteristic features of nearly all tropical climates (see Plates 25 and 26).

Following the evolution of this simple picture of tropical weather processes, attempts were made between 1920 and 1940 to introduce mid-latitude frontal concepts. Little real progress was made, however, as a result of the apparent general absence of significant air-mass contrasts. Furthermore, the small surface pressure gradients of most tropical disturbances (other than the hurricane variety) tend to go unnoticed due to the large semi-diurnal pressure oscillation in the tropics. Pressure varies by about 2–3 mb with maxima around 1000 and 2200 hours and minima at 0400 and 1600 hours. It must also be recalled that the wind direction provides no guide to the pressure pattern in low latitudes. The small Coriolis force prevents the wind from attaining geostrophic balance and consequently the techniques used for analyzing mid-latitude weather maps have to be abandoned.

During and after the Second World War more numerous weather observations revealed the inadequacy of the earlier views. It became evident that weather changes are frequent and complex with quite distinct types of weather systems in different tropical lands and with considerable climatic differences even over the ocean areas. It also became apparent that much smaller trigger mechanisms were necessary to produce disturbances in the flow of high-energy, tropical air masses than those associated with mid-latitude depressions, and yet, paradoxically, tropical cyclones are infrequent. The systems responsible for these contrasts are examined in the following sections. In this chapter it is intended to look first at the synoptic-scale disturbances and then at the Asian monsoon regime.

B. TROPICAL DISTURBANCES

Tropical depressions are rarely frontal; rather they are perturbations of the wind flow. Three major types may be recognized: the wave disturbances of

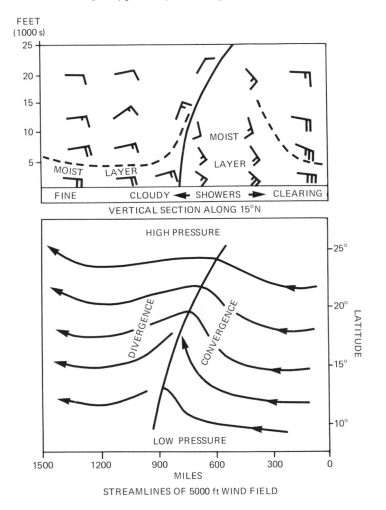

Fig. 6.1. Easterly wave model. The lower diagram shows the instantaneous wind pattern (streamlines) at about 1500-meters (5000 feet); the upper one a vertical section along latitude 15°N. The wind arrows show direction and speed at different levels; the dashed line represents the trade-wind inversion; the heavy line marks the wave trough (after Riehl and Malkus).

the trades and equatorial easterlies; cyclonic vortices (including the violent tropical cyclones); and linear systems.

1. Wave disturbances. The trades (see Fig. 3.21) and the equatorial easterlies of the central Pacific develop wave troughs which are scarcely detectable in the surface pressure field, but may form a closed low-pressure area in the middle troposphere. These systems are quite unlike mid-latitude depressions. In the *easterly wave* of the Caribbean area there is a weak pressure trough which usually slopes eastward with height (Fig. 6.1) and typically shows the main development of cumulonimbus cloud and thundery showers *behind* the trough line. This pattern is associated with horizontal and vertical motion in the easterlies. Behind the trough air undergoes convergence, while ahead of it there is divergence (Chapter 3, Section C.5). This follows from the equation for the conservation of potential vorticity (compare Chapter 4, Section F), which assumes that the air traveling at a given level does not change its potential temperature (or dry-adiabatic motion; see Chapter 2, Section E):

$$\frac{f + \zeta}{\Delta p} = k,$$

f = the Coriolis parameter, ζ = relative vorticity (cyclonic positive), and Δp = the depth of the tropospheric air column. Air overtaking the trough line is moving both polewards (f increasing) and towards a zone of cyclonic curvature (ζ increasing), so that if the left-hand side of the equation is to remain constant Δp must increase. This vertical expansion of the air column necessitates horizontal contraction (convergence). Conversely there is divergence in the air moving southward ahead of the trough and curving anticyclonically. The divergent zone is characterized by descending, drying air with only a shallow moist layer near the surface, while in the vicinity of the trough and behind it the moist layer may be 4500 meters (15,000 ft) or more deep.

The passage of such a transverse wave in the trades commonly produces the following weather sequence:

In the ridge ahead of the trough	fine weather, scattered cumulus cloud, some haze.
Close to the trough line	well-developed cumulus, occasional showers, improving visibility.
Behind the trough	veer of wind direction, heavy cumulus and cumulonimbus, moderate or heavy thundery showers, and a decrease of temperature.

Fig. 6.1 illustrates these features and shows that the dimensions of the wave are about half those of a mid-latitude depression. Their speed of movement is also slower, being 20–24 kmph (12–15 mph) on average in the Caribbean.

It is obviously very difficult to trace the beginnings of these waves over the tropical Atlantic, but the structure of the trades is known to be a basic cause. In the eastern sectors of the subtropical high-pressure cells active subsidence maintains a pronounced inversion at 450 to 600 meters (1500 to 2000 feet) (Fig. 6.2). Downstream the height of the inversion base rises (Fig. 6.3) because the subsidence decreases away from the eastern part of the anticyclone and cumulus towers penetrate through the inversion from time to time, spreading moisture into the dry air above. Easterly waves tend to develop in the Caribbean when the trade-wind inversion is weak or even

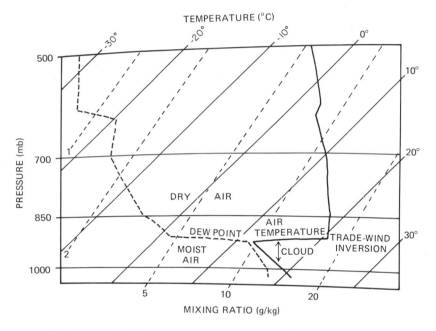

Fig. 6.2. The vertical structure of trade wind air at 30°N, 140°W at 0300 G M T on July 10, 1949. The mixing ratio is the saturation value (based on Riehl, 1954).

absent during summer and autumn, whereas in winter and spring accentuated subsidence aloft inhibits their growth. Another factor which may initiate waves in the easterlies is the penetration of cold fronts into low latitudes. This is common in the sector between two subtropical high-pressure cells where the equatorward portion of the front tends to fracture, generating a westward-moving wave.

The influence of these features on regional climate is illustrated by the rainfall regime. For example, there is a late summer maximum at Martinique (Fig. 6.4) in the Windward Islands (15°N) when subsidence is weak, although some of the autumn rainfall is associated with tropical storms. In many trade-wind areas the rainfall occurs in a few rainstorms associated with some

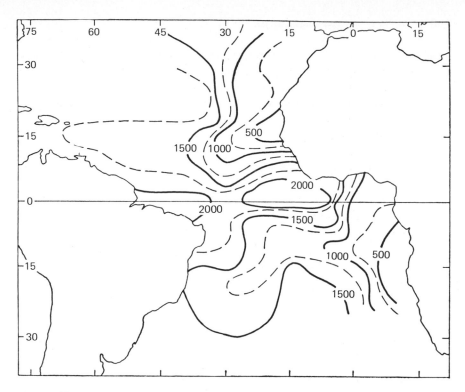

Fig. 6.3. The height (in meters) of the base of the trade-wind inversion over the tropical Atlantic (from Riehl, 1954).

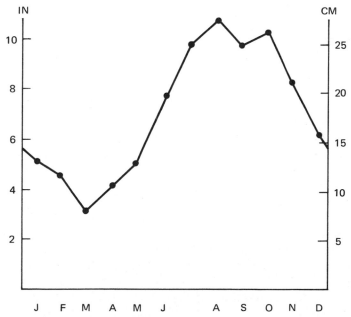

Fig. 6.4. Average monthly rainfall (cm left, inches right) at Fort de France, Martinique (based on Kendrew, 1953).

form of disturbance. Over a 10-year period Oahu (Hawaii) had an average of 24 rainstorms per year of which 10 accounted for more than two-thirds of the annual precipitation. There is quite high variability of rainfall from year to year in such areas, since a small reduction in the frequency of disturbances can have a large effect on rainfall totals.

In the central equatorial Pacific the trade-wind systems of the two hemispheres converge in the Equatorial Trough and wave disturbances may be generated if the trough is sufficiently removed from the equator (usually to the north) to provide a small Coriolis force to begin cyclone motion. These disturbances quite often become unstable forming a cyclonic vortex as they travel westwards towards the Philippines, but the winds do not necessarily

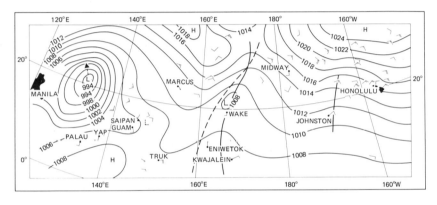

Fig. 6.5. The surface synoptic chart for part of the north-west Pacific on August 17, 1957. The movements of the central wave trough and of the closed circulation during the following 24 hours are shown by the dashed line and arrow, respectively. The dashed L just east of Saipan indicated the location in which another low pressure system subsequently developed. Plate 27 shows the cloud formation along the convergence zone just east of Wake Island (from Malkus and Riehl, 1964).

attain hurricane strength. The synoptic chart for part of the northwest Pacific on August 17, 1957 (Fig. 6.5) shows three developmental stages of tropical low pressure systems. An incipient easterly wave has formed west of Hawaii which, however, filled and dissipated during the next 24 hours. A well-developed wave is evident near Wake Island, having spectacular cumulus towers extending up to over 9100 meters (30,000 ft) along the convergence zone some 480 km (300 miles) east of it (Plate 27). This wave developed within 48 hours into a circular tropical storm with winds of up to 20 m/sec (46 mph), but not into a full hurricane. A strong, closed, northwest moving circulation is situated east of the Philippines. Equatorial waves may form on both sides of the equator in an easterly current located between about 5°N and S. In such cases divergence ahead of a trough in the northern hemisphere is paired with convergence behind a trough line located further to the west in the southern hemisphere. The reader may confirm that this should be so by applying the equation for the conservation of potential vorticity, remembering that both f and ζ operate in the reverse sense in the southern hemisphere.

2. Cyclones. (a) Hurricanes.

The most notorious type of cyclone is the tropical hurricane (or typhoon). Because these cause widespread damage over land areas and are a serious shipping hazard, considerable attention has been given to forecasting their development and movement so that their origin and structure are beginning to be understood. Naturally the catastrophic force of a hurricane makes it a very difficult phenomenon to investigate, but some assistance is now obtained from aircraft reconnaissance

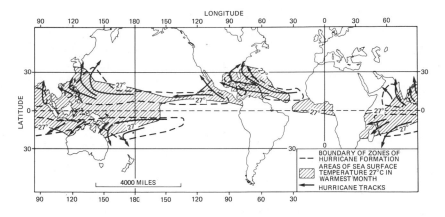

Fig. 6.6. Areas of hurricane formation, their principal tracks and mean sea-surface temperatures of the warmest month (after Palmén, 1948 and Bergeron, 1954).

flights sent out during the hurricane season, from radar observations of cloud and precipitation structure, and from satellite photography (see Plates 28 and 29).

The typical hurricane system has a diameter of about 650 km (400 miles), less than half that of a mid-latitude depression, although typhoons in the China Sea are often much larger. The central pressure at sea level is commonly 950 mb and exceptionally falls below 920 mb. Hurricane winds are defined arbitrarily as 33 m/sec (74 mph) or more and in many storms they exceed 50 m/sce (120 mph) in a ring some 10–60 km wide around the eye, while in the extensive outer zone winds are of gale force. The great vertical development of cumulonimbus clouds with tops at over 12,000 meters (40,000 ft) reflects the immense convective activity concentrated in such systems. Radar studies show that the convective cells are normally organized in bands which spiral inward towards the center.

A number of conditions are necessary, even if not always sufficient, for hurricane formation. One requirement as shown by Fig. 6.6 is an extensive ocean area with a surface temperature greater than 27°C (80°F). Cyclones rarely form within about 5° latitude of the equator, where the Coriolis parameter is close to zero, or in zones of strong vertical wind shear (for example, beneath a jet stream) as both factors inhibit the development of an organized vortex. There is also a definite connection between the seasonal position of the Equatorial Trough and zones of hurricane formation, which is borne out by the fact that no hurricanes occur in the South Atlantic (where the trough never lies south of 5°S) or in the southeast Pacific (where the trough remains north of the equator). On the other hand, recent satellite photographs over the northeast Pacific show an unexpected number of cyclonic vortices in summer, many of which move westwards near the trough line about 10°–15°N.

Hurricanes can also originate deep within the trades circulation; only 50% of Philippine storms, for example, form within 100 km (60 miles) of the Equatorial Trough. The development regions of hurricanes mainly lie over the western sections of the Atlantic, Pacific, and Indian Oceans where the subtropical high-pressure cells do not cause subsidence and stability.

The main hurricane (and typhoon) activity in the northern hemisphere is in late summer and autumn during the time of the Equatorial Trough's northward displacement. The hurricane season occurs in the western Atlantic mainly between July and October with a marked peak in September, and in the western Pacific between July and October. A small number of storms may affect both areas as early as May and as late as December. The late summer–autumn maximum is also found in the other areas, although there is a secondary, early summer maximum in the Bay of Bengal.

Annual frequencies of tropical storms and cyclones are summarized in Table 6.1. The figures are only approximate, since in some cases it may be uncertain as to whether or not the winds actually exceeded hurricane force and storms in the more remote parts of the South Pacific and Indian

Table 6.1: The mean annual frequency of tropical storms and cyclones (*after Ramage, Cry, Sircar, and others*).

	Tropical storms Winds $\geq$ 17 m/sec		Tropical cyclones Winds $\geq$ 33 m/sec	
Western North Pacific	22.0	(1884–1953)	19.4	(1924–53)
Eastern North Pacific	5.7	(1910–40)	2.2	(1910–40)
Western North Atlantic	7.9	(1901–63)	4.6	(1901–63)
Bay of Bengal	13.0	(1890–1950)	4.7	(1890–1950)
Arabian Sea	1.3	(1881–1937)	0.7	(1881–1937)
Southwest Indian Ocean			4.7	(1848–1935)
Southeast Indian Ocean			2.1	(1919–56)
Western South Pacific			4.0	(1940–56)

Oceans may have escaped detection prior to the use of weather satellites to track such storms.

Early theories of hurricane development held that convection cells generated a sudden and massive release of latent heat to provide energy for the storm. Although convection cells were regarded as an integral part of the hurricane system, their scale was thought to be too small for them to account for the growth of a storm hundreds of kilometers in diameter. Recent research, however, is modifying this picture considerably. Energy is apparently transferred from the cumulus-scale to the large-scale circulation of the storm through the organization of the clouds into spiral bands, although the nature of this process is still being investigated. There is now ample evidence to show that hurricanes form from preexisting disturbances, but while many of these disturbances develop as closed low-pressure cells few attain full hurricane intensity. The key to this problem appears to be the presence of an anticyclone in the upper troposphere. This is essential for high-level outflow (see Chapter 3, Sections C.4 and 5), which in turn allows the development of very low-pressure and high wind speeds near the surface. A distinctive feature of the hurricane is the warm vortex, since other tropical depressions and incipient storms have a cold core area of shower activity. The warm core develops through the action of 100–200 cumulonimbus towers releasing latent heat of condensation. Observations show that although these hot towers form less than 1% of the storm area within a radius of about 400 km (230 miles), their effect is sufficient to change the environment. The warm core is vital to hurricane growth because it intensifies the upper anticyclone, leading to a feedback effect by stimulating the low-level influx of heat and moisture which further intensifies convective activity, latent heat release, and therefore the upper-level high pressure. The thermally-direct circulation converts the heat increment into potential energy and a small fraction of this—about 3%—is transformed into kinetic energy.

In the *eye*, or innermost region of the storm, adiabatic warming of descending air accentuates the high temperatures (Fig. 6.7), although, since high temperatures are also observed in the eye-wall cloud masses, subsiding air can only be one contributory factor. The eye has a diameter of some 30–50 km (20–30 miles), within which the air is virtually calm and the cloud cover may be broken. The mechanics of the eye's inception are still largely unknown. If the rotating air conserved absolute angular momentum, wind speeds would become infinite at the center and clearly this is not the case. The strong winds surrounding the eye are more or less in cyclostrophic balance, with the small radial distance providing a large centripetal acceleration (see Chapter 3, Section A.4). The link between the horizontal air circulation and the vertical motion around the eye is complex. One idea, due to R. S. Scorer, is that the cumulonimbus anvils redistribute the angular momentum of the air as they spread out and stir their environment by

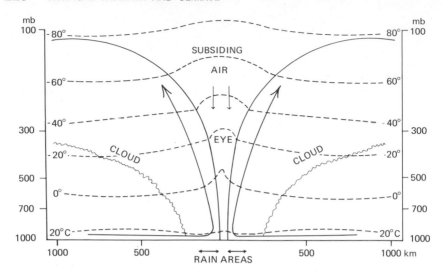

Fig. 6.7. A model of the vertical structure of a hurricane (after Palmén, 1948 and Bergeron 1954).

mixing with it, so reducing its rotation. The effect is to give a concentration of rotation near the center.

The supply of heat and moisture combined with low frictional drag at the sea surface, the release of latent heat through condensation, and the removal of the air aloft are essential conditions for the maintenance of hurricane intensity. As soon as one of these ingredients diminishes the storm decays. This can occur quite rapidly if the track (determined by the general upper tropospheric flow) takes the vortex over a cool sea surface or over land. In the latter case the increased friction hastens the process of filling while the cutting-off of the moisture supply removes one of the major sources of heat. Rapid decay also occurs when cold air is drawn into the circulation or when the upper-level divergence pattern moves away from the storm.

Hurricanes usually move at some 16–24 kmph. (10–15 mph), controlled primarily by the rate of movement of the upper warm core. Almost invariably they recurve poleward around the western margins of the subtropical high-pressure cells, entering the circulation of the westerlies, where they die out or degenerate into extratropical depressions. Some of these systems retain an intense circulation and the high winds and waves can still wreak havoc. This is not uncommon along the Atlantic coast of the United States and occasionally eastern Canada. Similarly, in the western North Pacific. recurved typhoons are a major element in the climate of Japan (see Chapter 6, Section C.4) and may occur in any month. There is an average frequency of 12 typhoons per year over southern Japan and neighboring sea areas.

To sum up. The hurricane develops from an initial disturbance which, under favorable environmental conditions, grows first into a tropical depression and then a tropical storm (with wind speeds of 17–33 m/sec or 39–73 mph). The tropical storm stage may persist 4–5 days, whereas the hurricane stage usually lasts for only 2–3 days. The main energy source is latent heat derived from condensed water vapor, and for this reason hurricanes are generated and continue to gather strength only within the confines of warm oceans. The cold-cored tropical storm is transformed into a warm-cored hurricane in association with the release of latent heat in cumulonimbus towers, and this establishes or intensifies an upper tropospheric anticyclone cell. Thus high-level outflow maintains the ascent and low-level inflow in order to provide a continual generation of potential energy (from latent heat) and the transformation of this into kinetic energy. The inner eye which forms during the cold-core tropical storm stage is an essential element in the life cycle.

(b) Other tropical depressions. Not all cyclonic systems in the tropics are of the intense hurricane variety. Another type of low which is of climatic importance is the *monsoon depression* which affects India in summer (see Fig. 6.19). There appear to be two distinct kinds of monsoon depression. One group, which affects the Bay of Bengal and central and northern India, appears to be similar in origin to the tropical cyclones already discussed in possessing a pronounced surface vortex and occasionally reaching full hurricane strength. Depressions of the second group are relatively weak near the surface but intense in the middle troposphere. They occur mainly in the northeastern Arabian Sea, though similar circulation systems are observed, especially during winter, in the eastern North Pacific where they have been termed *subtropical cyclones*. The latter are known to develop from the cutting-off in low latitudes of a cold upper-level wave in the westerlies (compare Chapter 4, Section F.4). They possess a broad eye of some 150 km (100 miles) in radius with little cloud, surrounded by a belt of cloud and precipitation about 300 km (200 miles) wide. Such cyclones are very persistent and tend to be reabsorbed eventually by a trough in the upper westerlies.

3. Linear systems. The early literature on tropical weather and climate made extensive use of the frontal concept. In particular, the term, Inter-Tropical Front (ITF), was widely applied to the boundary between the trade winds of the two hemispheres. Over continental areas such as West Africa and southern Asia, where very hot, dry continental tropical air meets cooler, humid equatorial air, this term has some limited applicability. Sharp temperature and humidity gradients exist here, but the frontal surface is seldom a conventional weather-producing mechanism like the Polar Front (see Fig. 6.18), and certainly the air-mass boundary in West Africa is south of the Equatorial Trough. Elsewhere in low latitudes true fronts (with a marked density contrast) are rare.

An important category of linear systems is the confluence line (or *asymptote of confluence*) which occurs when the streamlines of the wind field converge (see Chapter 3, Section C.5). The Equatorial Trough contains such an asymptote—the mean position of the ITCZ—due to the confluence of the trades. Where equatorial westerlies separate the trades, as in the sector between 60° and 160°E and in the eastern North Pacific, there are commonly two semipermanent confluence zones (see Chapter 3, Sections D.1 and 2). However, the ITCZ may be quite discontinuous spatially, as well as in time, and confluence lines are primarily located in wave disturbances and in cyclonic vortices (see Plate 30).

Two well-known linear systems are the *disturbance lines* of West Africa and the line squalls (known as *Sumatras*) of Malaya. The latter cross Malaya from the west in the early morning during the southwest monsoon and

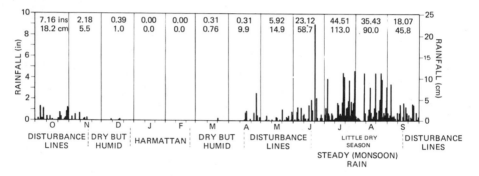

Fig. 6.8. Daily rainfall at Kortright, Freetown, Sierra Leone. October 1960–September 1961 (after Gregory, 1965).

appear to be initiated by the convergence effects of sea breezes in the Malacca Straits. The disturbance line is also a type of line squall but it is uncertain where and how they originate. They are several hundred kilometers long and travel westwards across West Africa at about 50 kmph (30 mph) giving squalls and thundery showers. Spring and autumn rainfall in West Africa is derived in large part from these disturbances. Fig. 6.8 for Kortright (Freetown, Sierra Leone) illustrates the daily rainfall amounts in 1960–1 associated with disturbance lines at 8°N. The summer monsoon rains make up the greater part of the total here, but their contribution diminishes northward. For example, in 1955 disturbance lines contributed about 30% of the annual rainfall over coastal Ghana and 90% in the north of the territory.

C. THE ASIAN MONSOON

The name *monsoon* is derived from the Arabic word (*mausim*) which means season, so explaining its application to large-scale seasonal reversals of the wind regime. The Asiatic seasonal reversal is notable for its immense extent and the penetration of its influence beyone tropical latitudes. For example, the surface circulation over China reflects this seasonal change:

	January	*July*
North China	60% of winds from W, NW, and N	57% of winds from SE, S, and SW
Southeast China	88% of winds from N, NE, and E	56% of winds from SE, S, and SW

However, such seasonal wind shifts at the surface are quite widespread and occur in many regions which would not traditionally be considered as monsoonal (Fig. 6.9). Although there is a rough accordance between these traditional regions and those experiencing over 60 percent frequency of the prevailing octant, it is obvious that a variety of unconnected mechanisms can produce significant seasonal wind shifts and that some more sophisticated mechanism must be sought for the more restricted tropical monsoonal circulations.

In summer the Equatorial Trough and the subtropical anticyclones are everywhere displaced northwards in response to the changing pattern of solar heating of the earth, and in southern Asia this movement is magnified by the effects of the land mass. However, the attractive simplicity of the traditional explanation, which envisages a monsoonal sea breeze directed towards a summer thermal low pressure over the continent, unfortunately provides an inadequate basis for understanding the workings of the system.

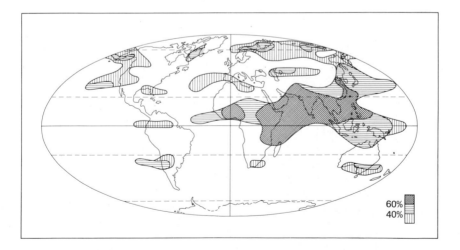

Fig. 6.9. Regions experiencing a seasonal surface wind shift of at least 120°, showing the frequency of the prevailing octant (after Chromov. Adapted from Flohn, 1960).

The Asiatic monsoon regime is a consequence of the interaction of both planetary and regional factors, both at the surface and in the upper troposphere. It is convenient to look at each season in turn.

1. Winter. Near the surface this is the season of the outblowing winter monsoon, but aloft the westerly airflow dominates. This, as we have seen, reflects the general pressure distribution. A shallow layer of cold, high-pressure air is centered over the continental interior, but this has disappeared even at 700 mb (see Fig. 3.11) where there is a trough over eastern Asia and zonal circulation over the continent. The upper westerlies split into two currents to the north and south of the high Tibetan plateau (Fig. 6.10), which exceeds 4000 meters (13,000 ft) over a vast area, to reunite again off the east coast of China (Fig. 6.11). These two branches have been attributed to the disruptive effect of the topographic barrier on the airflow, but the northern jet may be located far from the Tibetan plateau and two currents are also found to occur farther west where there is no obstacle to the flow. The branch over northern India corresponds with a strong latitudinal thermal gradient (from November to April) and it is probable that this factor, combined with the effect of the barrier to the north, is responsible for the anchoring of the southerly jet. This southern branch is the stronger, with an average speed of more than 66 m/sec (148 mph) at 200 mb, compared with about 20–25 m/sec (45–55 mph) in the northern one. Where the two unite over north China and south Japan the average velocity exceeds 66 m/sec (148 mph).

Air subsiding beneath this upper westerly current gives dry outblowing northerly winds from the subtropical anticyclone over northwestern India and West Pakistan. The surface wind direction is northwesterly over most of northern India, becoming northeasterly over Burma and East Pakistan and

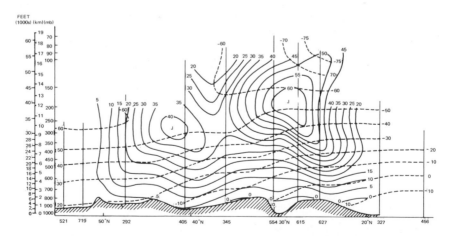

Fig. 6.10. The mean zonal geostrophic wind (in meters/sec; solid lines) and temperatures (in °C; dashed lines) along longitude 105°E for January–March 1956. The numbers along the base refer to upper-air observing stations (from Academica Sinica, 1957).

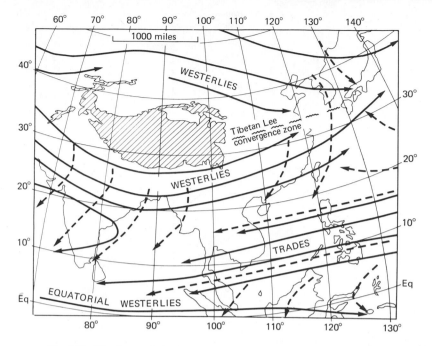

Fig. 6.11. The characteristic air circulation over southern and eastern Asia in winter (after Thompson, 1951; Flohn, 1960; Frost and Stephenson, 1965; and others). Solid lines indicate air flow at about 3000 meters (10,000 feet), and dashed lines that at about 600 meters (2000 feet). The names refer to the wind systems aloft.

easterly over peninsular India. Equally important is the steering of winter depressions over northern India by the upper jet. The lows, which are not usually frontal, appear to penetrate across the Middle East from the Mediterranean and are important sources of rainfall for northern India and West Pakistan (for example, Kalat, Fig. 6.12), especially as it falls when evaporation is at a minimum.

Some of these westerly depressions continue eastwards, redeveloping in the zone of jet-stream confluence about 30°N, 105°E over China, beyond the area of subsidence in the immediate lee of Tibet (Fig. 6.11), and it is significant that the mean axis of the winter jet stream over China shows a close correlation with the distribution of winter rainfall (Fig. 6.13). Other depressions affecting central and north China travel within the westerlies north of Tibet or are initiated by outbreaks of fresh cP air. In the rear of these depressions there are invasions of very cold air (for example, the *Buran* blizzards of Mongolia and Manchuria). The effect of such cold waves, comparable with the *northers* in the central and southern United States, is greatly to reduce mean temperatures. Winter mean temperatures in less-protected southern China are considerably below those at equivalent latitudes in India and, for example, temperatures at Calcutta and Hong Kong (both approximately

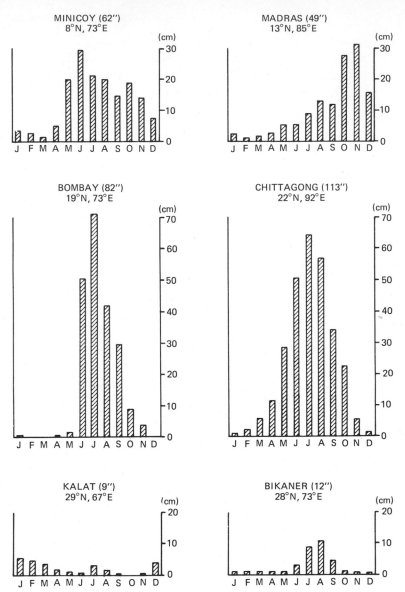

Fig. 6.12. Average monthly rainfall at six stations in the Indian region (based on CLIMAT, normals of the World Meteorological Organization for 1931–60). The annual total in inches is given after the station name.

$22\frac{1}{2}°$N) are, respectively, 19°C (67°F) and 16°C (60°F) in January and 22°C (71°F) and 15°C (59°F) in February.

2. Spring. The key to change during this transition season once again seems to be found in the pattern of the upper airflow. In March the upper westerlies begin their seasonal migration northwards, but whereas the northerly jet strengthens and begins to extend across central China and into

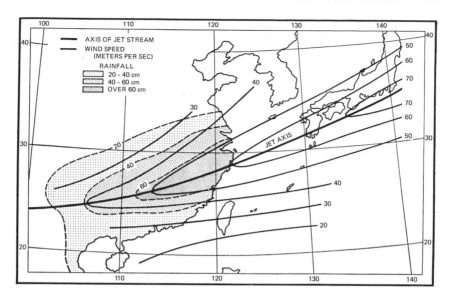

Fig. 6.13. The mean winter jet stream at 12 km over the Far East and the mean winter precipitation over China in cm (after Mohri and Yeh. From Trewartha, 1961).

Japan, the southerly branch remains positioned south of Tibet, although weakening in intensity.

The weather over northern India becomes hot, dry, and squally in response to the greater insolation. Mean temperatures at Delhi rise from 23°C (74°F) in March to 33°C (92°F) in May. The thermal low-pressure cell (see Chapter 4, Section G.2) reaches its maximum intensity at this time, but although onshore coastal winds develop the break of the monsoon rains is still a month away and other mechanisms cause only limited precipitation. Some precipitation occurs in the north with westerly disturbances, particularly towards the Ganges delta, where low-level inflow of warm, humid air triggers convective storms known as *nor'westers*. In the northwest, where less moisture is available, the convection generates violent squalls and dust storms termed *andhis*. The mechanism of these storms is not yet fully known, though high-level divergence ahead of waves in the subtropical westerly jet stream appears to be essential. The early onset of summer rains in Bengal, East Pakistan, Assam, and Burma (Chittagong, Fig. 6.12) is favored by an orographically produced trough in the westerlies at 300 mb, which is located about 85–90°E in May. Divergence ahead of this trough, combined with the convergence at low levels in maritime air from the Bay of Bengal, sets off thunder squalls. Another source of rainfall is the tropical disturbances in the Bay of Bengal, which have a secondary maximum of occurrence in April–May. Rainfall also occurs during this season over Ceylon and southern India (Minicoy, Fig. 6.9) in response to the northward movement of the Equatorial Trough and Equatorial Westerlies.

China has no equivalent of India's hot, premonsoon season. The low-level, northeasterly winter monsoon (reinforced by subsiding air from the upper westerlies) persists in north China and even in the south it only begins to be replaced by maritime tropical air in April–May. Thus, at Canton mean temperatures rise from only 17°C (63°F) in March to 27°C (80°F) in May, some 6°C (12°F) less than the mean values over northern India.

Westerly depressions are most frequent over China in spring. They form more readily over central Asia at this season as the continental anticyclone begins to weaken; also many develop in the jet-stream confluence zone. The average number crossing China per month during 1921–31 was as follows:

J	F	M	A	M	J	J	A	S	O	N	D	Year
7	8	9	11	10	8	5	3	3	6	7	7	86

Hence spring is wetter than winter and over most of central and southern China, the three months March–May contribute a quarter to a third of the annual rainfall.

3. Early summer. Generally during the last week in May the southern jet begins to break down, becoming first intermittent and then finally being diverted altogether to the north of the plateau. Over India the Equatorial Trough pushes northwards with each weakening of the upper westerlies south of Tibet, but the final burst of the monsoon, with the arrival of the humid, low-level southwesterlies, is not accomplished until the upper-air circulation has switched to its summer pattern (Fig. 6.14). The low-level changes are related to the establishment of a high-level easterly jet stream over southern Asia about 15°N. One theory suggests that this takes place in June as the col between the subtropical anticyclone cells of the west Pacific and the Arabian Sea at the 300-mb level is displaced northwestwards from a position about 15°N, 95°E in May towards central India. The northwestward movement of the monsoon (Fig. 6.15) is apparently related to the extension over India of the upper tropospheric easterlies.

The reorganization of the upper airflow has widespread effects in southern Asia. It is directly linked with the *Mai-yu* rains of China (which reach a peak about June 10–15), the onset of the southwest Indian monsoon, and the northerly retreat of the upper westerlies over the whole of the Middle East.

It must nevertheless be emphasized that it is still uncertain how far these changes are caused by events in the upper air or indeed whether the onset of the monsoon initiates a readjustment in the upper-air circulation. The presence of the Tibetan plateau is certainly of importance even if there is no significant barrier effect on the airflow. Heating of the plateau in early summer results in the formation of a thermal high-pressure cell (see Chapter

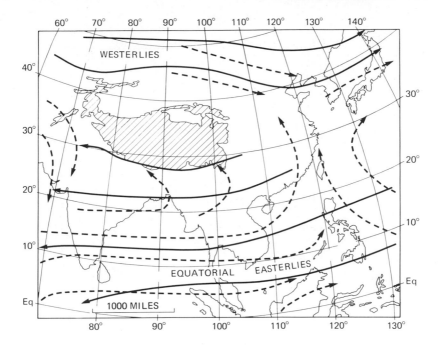

Fig. 6.14. The characteristic air circulation over southern and eastern Asia in summer (after Thompson, 1951; Flohn, 1960; Frost and Stephenson, 1965; and others). Solid lines indicate air flow at about 6000 meters (20,000 feet) and dashed lines that at about 600 meters (2000 feet). Note that the low-level flow is very uniform between about 600 and 3000 meters.

3, Section C.4) over the area at the level of its surface (about 600–500 mb) and, by producing easterly flow on the south side of the anticyclone, this undoubtedly assists the breakdown of the southern westerly jet. At the same time, the premonsoonal convective activity over the southeastern Tibetan Highlands provides a further heat source, by latent heat release, for the upper tropospheric anticyclone. Clearly, the interactions between the Indian monsoon regime and the Tibetan Highlands are decidedly complex.

Over China, the zonal westerlies retreat northwards in May–June and the westerly flow becomes concentrated north of the Tibetan plateau. The equatorial westerlies spread across southeast Asia from the Indian Ocean, giving a warm, humid air mass at least 3000 meters deep. Contrary to earlier views, the Pacific is only a moisture source when tropical southeasterlies extend westwards to affect the east coast. Most of the summer precipitation over China south of 30–35°N is derived from the northeastward extension of the south Asian monsoon current, seemingly in relation to surges in the flow causing zones of speed convergence (see Chapter 3, Section C.5). Central China is also affected by weak disturbances moving eastward along the Yangtze valley and by occasional cold fronts from the northwest.

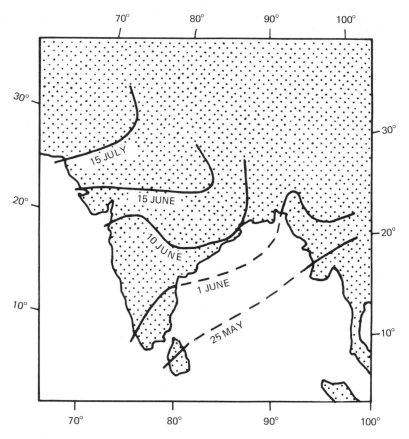

Fig. 6.15. The advance of the summer monsoon, indicated by average dates for the break of the monsoon (after Ananthakrishnan and Rajagopalachari, 1964. In Hutchings, 1964).

4. Summer. By mid-July monsoon air covers most of southern and south-eastern Asia and in India the Equatorial Trough is located about 25°N. North of the Tibetan plateau there is a rather weak upper westerly current with a (subtropical) high-pressure cell over the plateau. The southwest monsoon in southern Asia is overlain by strong upper easterlies with a pronounced jet at 150 mb (about 15 km or 50,000 ft) which extends westwards across South Arabia and Africa (Fig. 6.16). No easterly jets have so far been observed over the tropical Atlantic or Pacific. The jet is related to a steep lateral temperature gradient with the upper air getting progressively colder to the south. An important characteristic of the tropical easterly jet is the location of the main belt of summer rainfall on the right (or north) side of the axis upstream of the wind maximum and on the left side downstream, except for areas where the orographic effect is predominant (Fig. 6.16). The mean jet maximum is located about 15°N, 50–80°E.

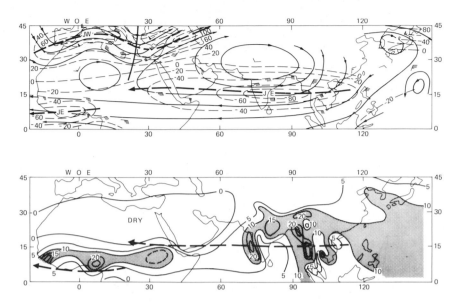

Fig. 6.16. The easterly tropical jet stream (from Koteswaram, 1958). Above. The location of the easterly jet streams at 200 mb on July 25, 1955. Streamlines are shown in solid lines and isotachs (wind speed) dashed. Wind speeds are given in knots (westerly components positive, easterly negative). Below. The average July rainfall (shaded areas receive more than 10 inches or 25 cm) in relation to the location of the easterly jet streams.

The monsoon current does not give rise to a simple pattern of weather over India, despite the fact that much of the country receives 80% or more of its annual precipitation during the monsoon season (Fig. 6.17). In the northwest a thin wedge of monsoon air is overlain by subsiding continental air (Fig. 6.18) and it is evident that the frontal surface is inactive in terms of weather. The inversion prevents convection and consequently little or no rain falls in the summer months in the arid northwest of the subcontinent (for example, Bikaner and Kalat, Fig. 6.12).

Around the head of the Bay of Bengal and along the Ganges valley the main weather mechanisms in summer are the monsoon depressions (see Chapter 6, Section B.2.*b*) which usually move westwards or northwestwards across India, steered by the upper easterlies. Fig. 6.19 shows that the main rainfall areas occur south of the Equatorial Trough, in association with the monsoon lows, and also on windward coasts and mountains of India, Burma, and Malaya. Without such disturbances the distribution of monsoon rains would be controlled to a much larger degree by orography.

It has recently been discovered that the southwesterly current over the Indian Ocean is quite dry near the equator, apart from a shallow moist layer at the surface. Moisture is acquired over the Arabian Sea, although even there an inversion indicates the presence of dry upper air originating

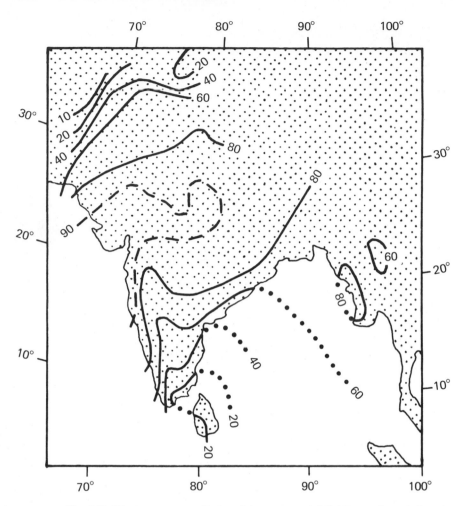

Fig. 6.17. The percentage contribution of the monsoon rainfall (June to September) to the annual total (after Rao and Ramamoorthy, 1960, in Indian Meteorological Department, 1960; and Ananthakrishnan and Rajagopalachari, 1964, in Hutchings, 1964).

perhaps over Arabia or East Africa. Convective instability is only released as the air slows down and converges at the coast and is forced over the Western Ghats. At Mangalore (13°N) there are on average 25 rain days per month in June, 28 in July, and 25 in August. The monthly rainfall averages are respectively 98 cm (38.6 in.), 106 cm (41.7 in.), and 58 cm (22.8 in.), accounting for 75% of the annual total. In the lee of the Ghats amounts are much reduced and there are semi-arid areas receiving less than 64 cm (25 in.) per year.

In southern India, excluding the southeast, there is a marked tendency for less rainfall when the Equatorial Trough is farthest north. Fig. 6.12

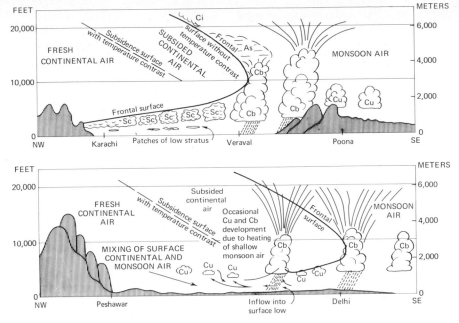

Fig. 6.18. The structure of the Inter-tropical Front over the north-western part of the Indian sub-continent (from Sawyer, 1947).

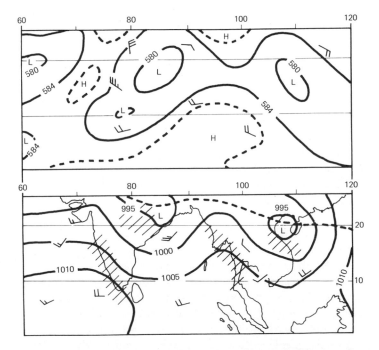

Fig. 6.19. Monsoon depressions at 1200 GMT, July 4, 1957. The upper diagram shows the height (geopotential decameters) of the 500-mb surface, the lower one the sea-level isobars. The broken lines represents the Equatorial Trough, and precipitation areas are shown by the oblique shading (based on the IGY charts of the Deutscher Wetterdienst).

shows a maximum at Minicoy in June with a secondary peak in October as the Equatorial Trough and its associated disturbances withdraw southward. This double peak occurs in much of interior peninsular India south of about 20°N and in western Ceylon, although autumn is the wettest period.

It is important to realize that the monsoon rains are highly variable from year-to-year, emphasizing the role played by disturbances in generating rainfall within the favorable environment created by the moist southwesterlies. *Breaks* occur in the monsoon rains when, during low-index periods, mid-latitude westerlies accompanied by the jet push southwards, weakening the Tibetan anticyclone or displacing it northeastwards. Westerly troughs travel along the southern edge of the Himalayas, giving heavy rain on the mountain slopes but little rain elsewhere. In part this may be due to the eastward extension across central India of the subtropical high over Arabia.

To the east over China, the surface airflow is southwesterly and the upper winds are weak with only a diffuse easterly current over southern China. According to traditional views the monsoon current reaches northern China by July. The annual rainfall regime shows a distinct summer maximum with, for example, 64% of the annual total occurring at Tientsin (39°N) in July and August. Nevertheless, much of the rain falls during thunderstorms associated with shallow lows and the existence of the ITCZ in this region is doubtful (see Fig. 4.15). The southerly winds, referred to above, which predominate over northern China in summer are not necessarily linked with the monsoon current farther south. Indeed this idea is the result of incorrect interpretation of streamline maps (of instantaneous airflow direction) as ones showing air trajectories (or the actual paths followed by air parcels). Cyclonic activity in northern China is attributable to the West Pacific Polar Front, forming between cP air and much-modified mT air.

In central and southern China the three summer months account for about 40–50% of the annual average precipitation, with another 30% or so being received in spring. In southeast China there is a rainfall singularity in the first half of July; a secondary minimum in the profile seems to result from the westward extension of the Pacific subtropical anticyclone over the coast of China.

A similar pattern of rainfall maxima occurs over southern and central Japan (Fig. 6.20), comprising two of the six natural seasons which have been recognized there. The main rains occur during the *Bai-u* season of the southeast monsoon resulting from waves, convergence zones, and closed circulations moving mainly in the tropical airstream round the Pacific subtropical anticyclone, but partly originating in a southwesterly stream which is the extension of the monsoon circulation of south Asia. The southeast circulation is displaced westwards from Japan by a zonal expansion of the subtropical anticyclone during late July and August giving a period of more settled sunny weather. The secondary precipitation maximum of the

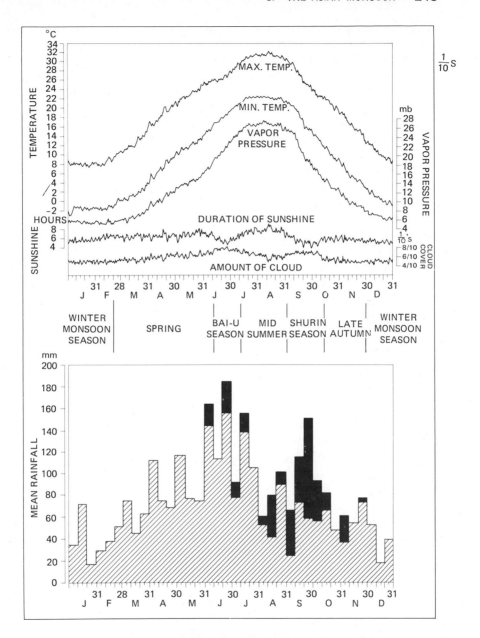

6.20. *Seasonal variation of daily normals at Nagoya, southern Japan (above), suggesting six natural seasons (from Maejima, 1967). Below are average 10-day precipitation amounts for a station in southern Japan, indicating in black the proportion of rainfall produced by typhoon circulations. The latter reaches a maximum during the Shurin season (after Saito, 1959. From Trewartha, 1961).*

Shurin season during September and early October coincides with an east-ward contraction of the Pacific subtropical anticyclone allowing low pressure systems and typhoons from the Pacific to swing north towards Japan. Although much of the Shurin rainfall is believed to be of typhoon origin (Fig. 6.20) some of it is undoubtedly associated with the southern sides of depressions moving along the southward-migrating Pacific Polar Front to the north, because there is a marked tendency for the autumn rains to begin first in the north of Japan and to spread southwards.

5. Autumn. Autumn sees the southward retreat of the Equatorial Trough and the break-up of the summer circulation systems. By October the easterly trades of the Pacific affect the Bay of Bengal at the 500-mb level and generate disturbances at their confluence with the equatorial westerlies. This is the major season for Bay of Bengal cyclones and it is these disturbances, rather than the onshore northeasterly monsoon, which cause the October–November maximum of rainfall in southeast India (for example, Madras, Fig. 6.12).

During October the westerly jet re-establishes itself south of the Tibetan plateau, often within a few days, and cool season conditions are restored over most of southern and eastern Asia.

D.　OTHER SOURCES OF CLIMATIC VARIATIONS IN THE TROPICS

The major systems of tropical weather and climate have now been discussed, yet various other elements help to create contrasts in tropical weather both in space and time.

1. Diurnal variations. Diurnal weather variations are most evident at coastal locations in the trade-wind belt. Land and sea breeze regimes (see Chapter 3, Section B.3) are well developed and daytime sea breezes may extend up to 1000–1200 meters (about 3000–4000 ft), with a heavy build-up of cumulus cloud and afternoon downpours. On large islands the sea breezes often converge towards the center so that an afternoon maximum of rainfall is observed. A typical case of afternoon maximum is illustrated by Fig. 6.21 for Nandi (Viti Levu, Fiji) in the southwest Pacific. The station has a lee exposure in both wet and dry season. This rainfall pattern is commonly believed to be widespread in the tropics, but over the open sea and on small islands a nighttime maximum (often with a peak near dawn) seems to occur. At Rarotonga (Fig. 6.21) 54% of the annual precipitation falls between 0800 and 2000. One theory is that the nocturnal radiative cooling of cloud tops makes the cloud layer less stable and encourages droplet growth by mixing of droplets at different temperatures (see Chapter 2, Section D). This effect would be at a maximum near dawn. Another factor is that the sea–air temperature difference, and consequently the oceanic heat supply to the atmosphere, is largest about 0300–0600 hours. Yet a further hypothesis suggests that the semidiurnal pressure oscillation en-

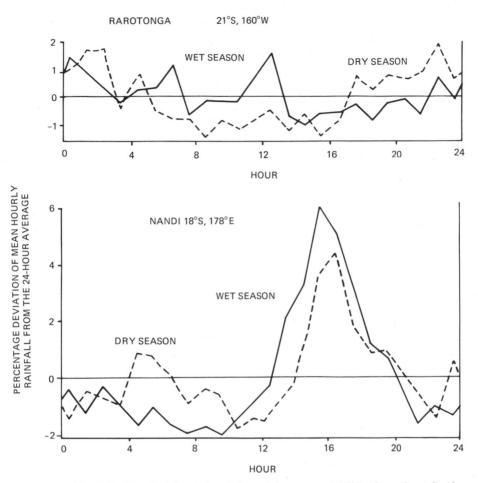

Fig. 6.21. The diurnal variation of wet and dry season rainfall in the southwest Pacific (after Finkelstein, 1964, in Hutchings, 1964). Amounts are shown as percentage deviations from the 24-hour average.

courages convergence and therefore convective activity in the early morning and evening, but divergence and suppression of convection around midday.

The Malayan peninsula displays very varied diurnal rainfall regimes in summer. The effects of land and sea breezes, anabatic and katabatic winds, and topography greatly complicate the rainfall pattern by their interactions with the low-level southwesterly monsoon current. For example, there is a nocturnal maximum in the Malacca Straits region associated with the convection set off by the convergence of land breezes from Malaya and Sumatra, whereas on the east coast of Malaya the maximum occurs in the late afternoon–early evening when sea breezes extend about 30 km inland against the monsoon southwesterlies, and convective cloud develops in the deeper sea breeze current over the coast strip. On the interior mountains the summer rains have an afternoon maximum due to the unhindered convection process.

2. Topographic effects. Relief and surface configuration have an especially marked effect on rainfall amounts in tropical regions where hot, humid air masses are frequent. In the Hawaiian Islands the mean annual total exceeds 760 cm (300 in.) on the mountains, with the world's largest mean annual total of 1199 cm (472 in.) on Mt Waialeale (Kauai), but land on the lee side suffers correspondingly accentuated sheltering effects with less than 50 cm (20 in.) over wide areas. On Hawaii itself the maximum falls on the eastern slopes at about 900 meters, while the 4200-meter summits of Mauna Loa and Mauna Kea, which rise above the trade-wind inversion, and their western slopes receive only 25–50 cm (10–20 in.). On the Hawaiian island of Oahu, the maximum precipitation occurs on the western slopes, just leeward of the 850-meter (2785-foot) summit with respect to the easterly trade winds. The following measurements in the Koolau Mountains, Oahu, show that the orographic factor is pronounced during summer when precipitation is associated with the easterlies, but in winter when precipitation is from cyclonic disturbances it is more evenly distributed:

| | | | | Source of Rainfall | |
| Location | Elevation | | Trade Winds | Cyclonic disturbances | |
	meters	feet	May 28– Sept. 3, 1957	Jan. 2–28, 1957	Mar. 5–6, 1957
Summit	850	(2785)	71.3 cm (28.09 in.)	49.9 cm (19.63 in.)	32.9 cm (12.94 in.)
760 meters (2500 ft) west of summit	625	(2050)	121.0 cm (47.64 in)	54.4 cm (21.41 in)	37.0 cm (14.56 in.)
7600 meters (25,000 ft) west of summit	350	(1150)	39.9 cm (12.97 in.)	46.7 cm (18.38 in.)	33.4 cm (13.16 in.)

(*After Mink, 1960.*)

The Khasi Hills in Assam are an exceptional instance of the combined effect of relief and surface configuration. Part of the monsoon current from the head of the Bay of Bengal (Fig. 6.14) is channeled by the topography towards the high ground and the sharp ascent, which follows the convergence of the airstream in the funnel-shaped lowland to the south, results in some of the heaviest annual rainfall totals recorded anywhere. Cherrapunji, at an elevation of 1340 meters, has a mean annual precipitation of 1144 cm (451 in.) and the extremes recorded there include 569 cm (224 in.) in July and 2299 cm (905 in.) in a calendar year.

Really high relief produces major changes in the main weather character-istics and is best treated as a special climatic type. The Kenya plateau, situated on the equator, has an average elevation of about 1500 meters, above which rise the three volcanic peaks of Mt Kilimanjaro (5800 meters), Mt Kenya (5200 meters), and Ruwenzori (5200 meters) nourishing per-manent glaciers above 4270 meters. Annual precipitation on the summit of Mt Kenya is about 114 cm (44 in.), similar to amounts on the plateau to the south, but on the southern slopes between 2100 and 3000 meters and on the eastern slopes between about 1400 and 2400 meters totals exceed 250 cm (100 in.). Kabete (at an elevation of 1800 meters near Nairobi) exhibits many of the features of tropical mountain climates, having a small annual temperature range (mean monthly temperatures are 19°C (67°F) for February and 16°C (60°F) for July), a high diurnal temperature range (averaging 9.5°C (17°F) in July and 13°C (24°F) in February) and a large average cloud cover (mean 7–8/10ths).

3. Cool ocean currents. Between the western coasts of the continents and the eastern rims of the subtropical high-pressure cells the ocean surface is relatively cold (see Fig. 3.29). This is a result of the importation of water from higher latitudes by the dominant currents and the slow upwelling (sometimes at the rate of about 1 meter in 24 hours) of water from inter-mediate depths due to the Ekman effect (Chapter 3, Section E.3) and to the coastal divergence (Chapter 3, Section C.5). This concentration of cold water gently cools the local air to dew point. As a result dry, warm air degenerates into relatively cool, clammy, foggy atmosphere with a compara-tively low temperature and little range along the west coast of South America between latitudes 4°S and 31°S, and off Southwest Africa (8°S to 32°S). Thus Lima (Peru, elevation 111 meters) has a mean annual temperature of only 18.3°C (65°F) and a mean annual temperature range of only 6.8°C (12°F). Callao, on the Peruvian coast, has a similarly low mean annual temperature of 19.4°C (67°F), whereas Bahia (at the same latitude on the Brazilian coast) has a corresponding figure of 25°C (77°F).

Nor is this cooling effect limited to stations near the coasts, for it is carried inland during the day at all times of the year by a pronounced sea-breeze effect (Chapter 3, Section B.3). Along the west coasts of South America and southwest Africa the sheltering effect from the dynamically stable easterly trades aloft provided by the nearby Andes and Namib Escarpment, respec-tively, allows incursions of shallow tongues of cold air to roll in from the southwest. These tongues of air are capped by strong inversions at between 600 to 1500 meters (2000–5000 ft) reinforcing the regionally low trade-wind inversions (see Fig. 6.3) and thereby precluding the development of strong convective cells, except where there is orographically-forced ascent. Thus, although the cool maritime air perpetually bathes the lower western slopes of the Andes in mist and low stratus cloud and Swakopmund (South West

Africa) has an average of 150 foggy days a year, little rain falls on the coastal lowlands. Lima has a total mean annual precipitation of only 4.6 cm (1.8 in.), although it receives frequent drizzle during the winter months of June to September, and Swakopmund has a mean annual rainfall of 1.6 cm (0.65 in.). Heavier rain occurs on the rare instances when large-scale pressure changes cause a cessation of the diurnal sea breeze or when modified air from the South Atlantic or South Indian Ocean is able to cross the continents at a time when the normal dynamic stability of the trade-wind flow is disturbed. In South West Africa the inversion is most likely to break down during either October or April allowing convectional storms to form, and Swakopmund recorded some 5.1 cm (2 in.) of rain on a single day in 1934. Under normal conditions, however, the occurrence of precipitation is mainly limited to the higher seaward mountain slopes. From central Colombia to northern Peru the diurnal tide of cold air rolls inland for some 60 km (38 miles), rising up the seaward slopes of the Western Cordillera and overflowing into the longitudinal Andean valleys like water over a weir (Fig. 6.22). Where this flow ascends or banks up against west-facing slopes it may under suitable

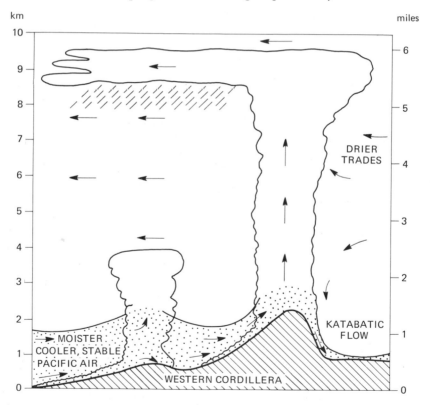

Fig. 6.22. The structure of the sea breeze in western Colombia (after Lopez. From Fairbridge, 1967).

conditions trigger off convectional instability in the overlying trades and produce thunderstorms. In South West Africa, however, the 'tide' flows inland for some 130 km (80 miles) and rises up the 6000-ft (1800-meter) Namib Escarpment without producing much rain because convectional instability is not produced and the adiabatic cooling of the air is more than offset by the radiational heating from the warm ground.

7

Urban and Forest Climates

The effects of climate upon all organisms and especially upon man have been recurrent themes in geography, but the growth of ideas relating to the wide field of ecology has brought about an increasing awareness of the influence exerted by the biosphere on weather and climate. The most extreme floristic environment—the forest—and the unintentional effects of man on local climate through his urban and industrial environments are considered in this chapter. However, since increasing interest is being shown in such topics as those relating to the possibility of weather modification and control, we begin with a brief résumé of some progress in applied meteorology.

On the smallest scale, man from earliest times has been concerned with modifying his most immediate environment to produce and sustain bodily comfort. Although advancing technology has made these efforts strikingly successful, their geographical significance lies both in the spatial variety of man's dwellings and in the economic cost of maintaining bodily comfort (Fig. 7.1) which, for modern advanced societies in extremely hot or cold conditions, can be very high. For example, Sweden's winter bill is estimated at 5% of the national income, of which about one half is spent on the insulation and heating of buildings.

Meteorological modifications by man involve temporary interventions with respect to short-term weather systems or conditions, and here some progress has also been made. Cold fogs can be locally dissipated by the use of dry ice or the release of propane gas through expansion nozzles to produce freezing. Warm fogs (that is, having drops above freezing temperatures) present bigger problems, but attempts at dissipation have shown some limited success in evaporating droplets by artificial heating, the use of large fans to draw down dry air from above, the sweeping out of fog particles by jets of water, and the injection of electrical charges into the fog to produce coagula-

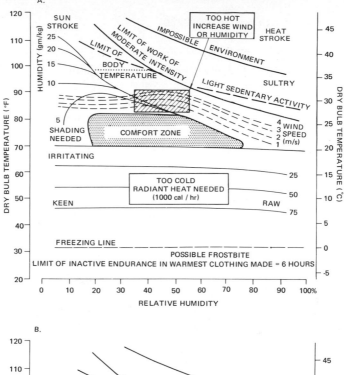

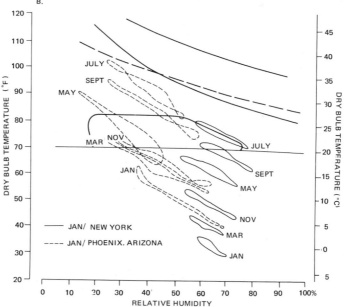

Fig. 7.1. The human relevance of climatic ranges (after Olgyay, 1963). A. Atmospheric comfort, discomfort and danger for inhabitants of temperate climatic zones. Within the stippled limits of dry-bulb temperature and relative humidity the body feels comfortable at elevations of less than 300 meters, with customary indoor clothing, and performing sedentary or light work. Outside these limits corrective measures are necessary to restore the feeling of comfort. Below the comfort zone radiant heat is needed, above it the skin temperature can be lowered by increasing the evaporation and heat convection either by raising the wind speed (for humid air) or by an increase of atmospheric moisture (for dry air). B. Mean daily dry-bulb temperatures and relative humidities during alternate months for New York City and Phoenix, Arizona, plotted on the comfort chart, indicating the need for both central heating and air conditioning.

251

tion. Attempts to produce precipitation have already been described in Chapter 2, Section D.1, and all that can be claimed is that the seeding of some cumulus clouds at temperatures of $-10°C$ to $-15°C$ with dry ice or silver iodide probably produces a mean increase of precipitation of 10 to 15% from clouds which are already precipitating or are "about to precipitate," with comparable increases up to 250 km downwind, whereas increases of up to 10% have resulted from seeding winter orographic storms. However, the seeding of depressions has produced no apparent precipitation increases, and it appears likely that clouds with an abundance of natural ice crystals or with above-freezing temperatures throughout are not susceptible to *rain-making*. The Russians have claimed some success in dissipating damaging hailstorms by the use of radar-directed artillery shells and rockets to inject silver iodide into high-liquid-water-content portions of clouds which freezes the available supercooled water, so preventing it from accreting as shells on growing ice crystals. Attempts to drain off lightning charges by seeding clouds with silver iodide or with millions of metallic needles have produced even less certain results. Of all current attempts by man to control meteorological events none are more important than those relating to hurricanes. There is some indication that the seeding of the rising air in the cumulus eye-wall may widen the ring of condensation and updraught, decrease the angular momentum of the storm, and thus the maximum speed of the winds. Such attempts are still in their infancy, as are plans to cut off surface evaporation ahead of hurricanes by spreading the ocean surface with oily materials.

Climatic modifications by man, the long-term continuation of large-scale meteorological controls, are for the most part of a highly speculative character. They vary from suggestions to produce widespread cirrus covers by rocket or aircraft seeding, to plans for locally changing sea temperatures and altering the surface patterns of insolation receipt by putting huge quantities of dust or metallic needles into orbit high above the earth. A particularly interesting idea is the creation of "thermal mountains" by painting desert surfaces black to decrease the albedo, stimulate convection, and thereby increase cloudiness and precipitation downwind. Another suggestion is to create major inland seas in arid basins with interior drainage, such as Lake Eyre, Australia, and thereby modify the moisture balance. The proposed use of nuclear energy and, in particular, thermonuclear explosions in large-scale *geographical engineering* projects also has possible climatic implications. The excavation of huge lakes, the production of open cuts to divert water through mountains, the removal of pack-ice, the melting of ice-caps, the truncation of mountain ranges, and the diversion of ocean currents (that is, by blocking the Bering Straits) have all been seriously proposed, but both the energy (Table 7.1) and expenditure involved and, particularly, the unknown attendant dangers in such permanent large-scale tampering with the earth's surface and atmosphere will postpone such schemes—perhaps permanently!

Table 7.1: Total energy of various individual phenomena and localized processes in the atmosphere (*from Sellers, 1965*). [Rates are relative to total solar energy intercepted by the earth (3.67×10^{21} cal/day)].

Solar energy received per day	1
Melting of average winter snow during the spring season	10^{-1}
Monsoon circulation	10^{-2}
World use of energy in 1950	10^{-2}
Strong earthquake	10^{-2}
Average depression	10^{-3}
Average hurricane	10^{-4}
Krakatoa explosion of August 1883	10^{-5}
Detonation of thermonuclear weapon in April 1954	10^{-5}
Kinetic energy of the general circulation	10^{-5}
Average squall-line	10^{-6}
Average magnetic storm	10^{-7}
Average summer thunderstorm	10^{-8}
Detonation of Nagasaki bomb in August 1945	10^{-8}
Average earthquake	10^{-8}
Burning of 7000 tons of coal	10^{-8}
Daily output of Hoover Dam	10^{-8}
Moderate rain (10 mm over Washington, D.C.)	10^{-8}
Average forest fire in the United States, 1952–3	10^{-9}
Average local shower	10^{-10}
Average tornado	10^{-11}
Street lighting on average night in New York City	10^{-11}
Average lightning stroke	10^{-13}
Average dust devil	10^{-15}
Individual gust near the earth's surface	10^{-17}
Meteorite	10^{-18}

The first step must be to develop theoretical mathematical models simulating the behavior of the earth-atmosphere system so that all the possible effects of these schemes can be predicted in advance. The development and operation of these models must await the collection of more data (largely from satellites) relating to the earth-atmosphere thermal and radiative properties, albedo, roughness, moisture content, evaporation, etc., and the development of larger computers to handle this mass of data.

A. URBAN CLIMATES

Until man develops direct methods of controlling the mechanisms of meteorology, his most profound climatic influence will remain the modification of local climates by the construction of large cities. The construction of every house, road, or factory destroys existing microclimates and creates new ones of great complexity depending on the design, density, and function of the building. Despite the great internal variation of urban climatic influences, it is possible to make certain generalizations regarding the effects of urban structures under three main headings:

(1) Heat production,
(2) Modification of atmospheric composition,
(3) Alteration of surface configuration and roughness.

1. Urban heat production. The thermal characteristics of urban areas are in marked contrast to those of the surrounding countryside. The three factors responsible for this are the direct production of heat by combustion; the gradual release of heat stored during the day by the brickwork, concrete, and similar materials; and the reflection from the atmospheric pollution layer of outgoing radiation back to the surface. The third factor is of rather minor importance.

It has been estimated that large German cities produce 15–30 cal/cm²/day from combustion, which compares with the amounts received by direct insolation of 50 cal/cm²/day in December and more than 500 cal/cm²/day in June. In Hamburg, prior to 1956, the average heat supply from coal burning during December was 40 cal/cm²/day, compared with a radiation from the sun and sky of 34 cal/cm²/day. Calculations for London suggest that winter domestic fuel consumption should not produce a temperature increase exceeding about 0.6°C in the city, whereas city temperatures are generally some 1°–2°C (2°–3°F) higher than those of the adjacent rural areas. The center of London had a mean annual temperature of 11.0°C (51.8°F) for the period 1931–60, comparing with 10.3°C (50.5°F) for the suburbs, and 9.6°C (49.2°F) for the surrounding countryside. Differences are even more marked during still-air conditions, especially at night (Fig. 7.2). For this *heat island* effect to operate effectively there must be wind speeds of less than 5–6 m/sec (12–14 mph), and it is especially in evidence on calm nights during summer and early autumn when it has steep cliff-like margins and the highest temperatures associated with the highest density of urban dwellings (Fig. 7.2). Thus the thermal contrasts of a city, like those of many other of its climatic features, depend on its topographic situation, and are greatest for sheltered sites with light winds. The fact that for London urban/rural temperature differences are greatest in summer, when direct heat combustion and atmospheric pollution are at a minimum, indicates that heat loss from buildings by radiation is the most important single factor contributing to the heat island effect. Seasonal differences are not necessarily the same, however, in other macroclimatic zones.

The effects on minimum temperatures are especially significant. Cologne, for example, has an average of 34% less days with minima below 0°C (32°F) than its surrounding area, the corresponding figure for Basle being 25% less. In London, Kew has an average of some 72 days with frost-free screen temperatures more than rural Wisley. Precipitation characteristics are also

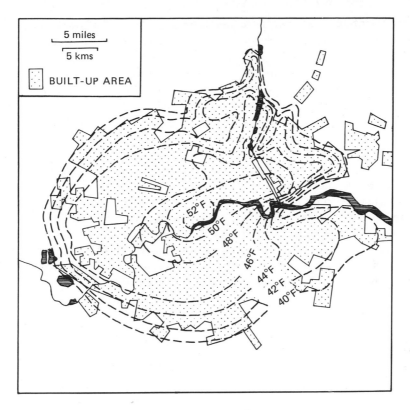

Fig. 7.2. Distribution of minimum temperatures (°F) in London on May 14, 1959, showing the relationship between the urban heat island and the built-up area (after Chandler, 1965).

affected and in pre-1917 Berlin 21% of the incidences of rural snowfall were associated with either sleet or rain in the city center.

Although it is difficult to isolate changes in temperatures which are due to urban controls from those due to other influences (see Chapter 8), it has been suggested that city growth is often accompanied by an increase in mean annual temperature, that of Osaka, Japan, rising by 2.6°C (4.5°F) in the last 100 years, and that of Tokyo by 1.5°C (3°F). These results may be coincidental, however, since there appears to be no linear relationship between city size and intensity of the heat island. Chandler has shown that Leicester, with a population of 270,000, exhibits warming comparable in intensity with that of central London over smaller sectors. The primary control of a city's thermal climate is the density of urban surfaces, that is, the total *surface* area of buildings and roads, and the building geometry. The vertical extent of the heat island is little known, although this is now a subject of study. In the case of cities with skyscrapers the vertical and horizontal patterns of wind and temperature must be very complex.

2. Modification of atmospheric composition. The city atmosphere is notoriously liable to pollution, particularly as the result of combustion, by smoke, dust, sulphur dioxide, and other gases. These have the effect of changing the thermal properties of the atmosphere, cutting down the passage of sunlight, and providing abundant condensation nuclei. The concentration of such nuclei ($0.01-0.1\mu$ diameter) averages some $9500/cm^3$ in the British countryside, but is typically $150,000/cm^3$ and can reach $4,000,000/cm^3$ in cities. Similarly, the concentration of dust particles ($0.5-10\mu$ diameter) has been measured at $25-30/cm^3$ in the city of Leipzig, as against $1-2/cm^3$ in the rural environs. The sources of pollution are many, and it has been estimated that, whereas 80–90% of London's smoke is produced by domestic fires, these are responsible for only 30% of the sulphur dioxide—the remainder being contributed by electricity power stations (41%) and factories (29%). It is important to note that clean air legislation has had the effect of decreasing the total smoke emission in London of 1.4×10^8 kg (141,000 tons) in 1952 to 0.9×10^8 kg (89,000 tons) in 1960. The greatest concentrations of smoke generally occur with low wind speeds, low vertical turbulence, temperature

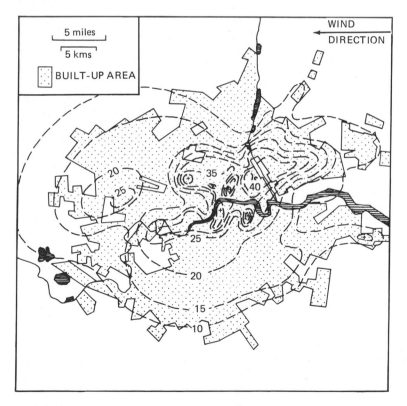

Fig. 7.3. Mean smoke concentration ($mg/100m^3$) with easterly winds in London for the period October 1958 to March 1959 (after Chandler, 1965).

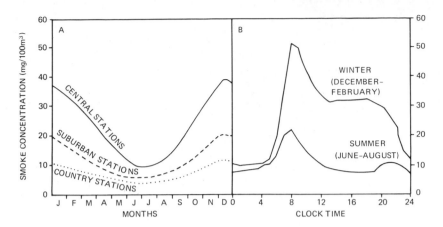

Fig. 7.4. Annual (A) and daily (B) cycles of smoke pollution at Leicester for the period 1937–9 (after Meetham, 1952).

inversions, high relative humidities, and air moving from the pollution sources of factory districts or areas of high-density housing (Fig. 7.3 and Plate 31). The character of domestic heating and power demands causes city smoke pollution to take on striking seasonal and diurnal cycles, with the greatest concentrations occurring about 0800 hours in December (Fig. 7.4B). The sudden morning increase is also partly a result of natural processes. Pollution trapped during the night beneath a stable layer a few hundred meters above the surface may be brought back to ground level when thermal convection sets off vertical mixing. This process is known as *fumigation*.

The most direct effect of atmospheric pollution is to reduce incoming radiation and sunshine. Pollution, plus the associated fog (termed "smog"), causes some British cities to lose 25–55% of the incoming solar radiation during the period November to March. In 1945 it was estimated that the city of Leicester lost 30% of incoming radiation in winter, as against 6% in summer. These losses are naturally greatest when the sun's rays strike the smog layer at low angle. Compared with the radiation received in the surrounding countryside, Vienna loses 15–21% of radiation when the sun's altitude is 30°, but the loss rises to 29–36% with an altitude of 10°. For London, the striking decrease of sunshine between the suburbs and the city center (Fig. 7.5) can mean a loss of mean daily sunshine of 16 minutes in the outer suburbs, 25 minutes in the inner suburbs, and 44 minutes in the city center. It must be remembered, however, that smog layers also impede the reradiation of surface heat at night, and that this blanketing effect contributes to the higher nighttime city temperatures.

The abundance of condensation nuclei in city atmospheres, particularly those situated on low-lying land adjacent to large rivers, explains the abundance of city fogs. During the period August 1944 to December 1946, for example, suburban Greenwich had a monthly average of more than 20 days

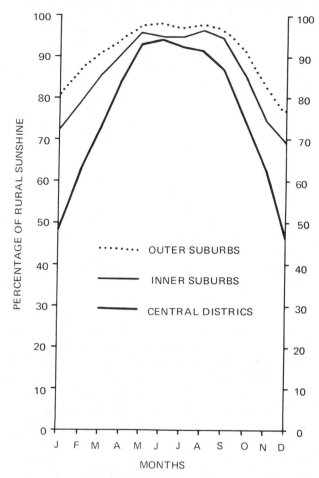

Fig. 7.5. Mean monthly bright sunshine recorded at three types of London stations for the years 1921–50, expressed as a percentage of that in adjacent rural areas (after Chandler, 1965).

with good visibility at 0900 hours, whereas central London had less than 15. Occasionally very stable atmospheric conditions combine with excessive pollution production to give dense smog of a lethal character. During the period December 5–9, 1952 a temperature inversion over London caused a dense fog with visibility less than 10 meters for 48 consecutive hours, resulting in 12,000 more deaths (mainly from chest complaints) during the period December 1952 to February 1953 compared with the same period the previous year. Whereas in London smog is a winter phenomenon associated with a radiation inversion, in Los Angeles smog occurs during the daytime in summer and autumn. In this case the inversion is commonly due to subsidence and the pollution haze is caused by car exhaust gases (here 3.7 million cars produce over 12,000 tons of pollutant daily). Recognition of this health hazard is leading to stricter pollution control by many metropolitan areas and in Tokyo citizens sometimes wear respiratory masks on the streets in self-defense!

The close association of the incidence of fog with increasing industrialization and urbanization is well shown by the city of Prague, where the mean annual number of days with fog rose from 79 during the period 1860–80 to 217 during 1900–20. Less sure is the effect of pollution on cloud cover and precipitation, although mid-latitude cities are believed to be 5–10% more cloudy than surrounding rural areas, and the mean annual precipitation of the town of Rochdale, Lancashire, is held to have increased during its industrial growth from 108.7 cm (42.81 in.) in 1898–1907 to 123.6 cm (48.65 in.) in 1918–27. In the latter instance the greatest increases appear to have occurred on weekdays, when the factories were working! Even so, it is quite possible that such differences represent chance fluctuations or widespread climatic trends.

3. Modification of surface characteristics. Urban structures have considerable effects on the movement of air both by producing turbulence as a result of their *roughening* the surface and by the channeling effects of the street gorges. On the average, city wind speeds are lower than those recorded in the surrounding open country owing to the sheltering effect of the buildings, and central average wind speeds are usually at least 5% less than those of the suburbs. In 1935, for example, winds exceeding 10.5 m/sec (24 mph) were recorded on the relatively open Croydon Airport (London suburbs) for a total of 371 hours, whereas the corresponding figure was only 13 hours for the closely built-up South Kensington. However, the urban effect on air motion varies greatly depending on the time of day and season. During the day city wind speeds are considerably less than those of surrounding rural areas, but during the night the greater mechanical turbulence over the city means that the higher wind speeds aloft are transferred to the air at lower levels by turbulent mixing. During the day (1300 hours) the mean annual wind speed for London Airport (open country within the suburbs was) 2.9 m/sec (6.4 mph), compared with 2.1 m/sec (4.7 mph) in central London for the period 1961–62. The comparative figures for night (0100 hours) were 2.2 m/sec (4.9 mph) and 2.5 m/sec (5.6 mph). Rural–urban wind speed differences are most marked with strong winds and the effects are therefore more evident during winter than during summer, when a higher proportion of low speeds is recorded in temperate latitudes.

The effect of urbanization on surface moisture relationships is also important. The absence of large bodies of standing water and the rapid removal of surface runoff through drains decreases local evaporation. In addition, the lack of an extensive vegetational cover eliminates much evapotranspiration. For these reasons the air of mid-latitude cities has a tendency towards lower

absolute humidities than that of their surroundings, especially under conditions of light wind and cloudy skies. On other occasions of calm, clear
weather the streets trap warm air which retains its moisture because less dew
is deposited on the warm surfaces of the city. Humidity contrasts between
urban and rural areas are most noticeable in the case of relative humidity,
which can be as much as 30% less in the city by night as a result of the higher
temperatures. Urban influences on precipitation (excluding fog) are much
more difficult to determine with any precision, partly because there are few
rain gauges in cities. It is now fairly certain, however, that urban areas in
Europe and North America are responsible for local conditions which, in
summer especially, can trigger-off excesses of precipitation under marginal
conditions (Fig. 7.6). These triggers involve the orographic and turbulence
effects of the buildings, the increased density of condensation nuclei, and
thermal convection. Recordings for Munich showed 11% more days of light
rain (0.1–0.5 mm or 0.004–0.02 in.) than in the surrounding countryside,
and Nürnberg has recorded 14% more thunderstorms than its rural environs.
European and North American cities generally record 11–18% more days
with light rain and thunder than their surrounding regions, responsible for

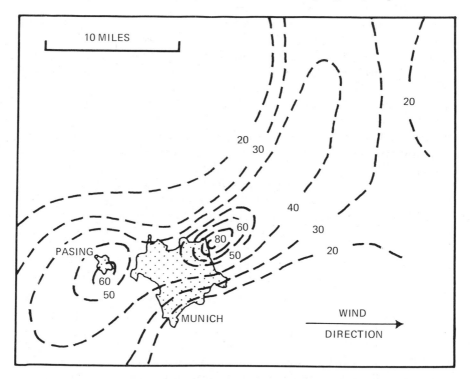

*Fig. 7.6. Distribution of precipitation (mm) from a cloudburst over Munich on July 25,
1929, which produced a secondary maximum over the suburb of Pasing (after J. Haeuser, and
Geiger, 1965).*

an increased urban precipitation of as much as 5–10%. On the other hand, in the majority of cases of medium-intensity frontal precipitation urban effects seem to be negligible.

B. FOREST CLIMATES

Geographers have been accustomed to considering the effects of climate on the surface vegetational cover, but it must be recognized that any variation of surface characteristics affects the local climate to some extent. Indeed, much research is currently being directed towards the microclimatic conditions associated with growing crops. When vegetation assumes its most dense and elevated forms in forests, the effects on local, and sometimes regional, climates are extremely important.

The vertical structure of a forest conditions, to a great extent, the control which it will exercise over local atmospheric conditions. At the same time it must be recognized that botanical differences are usually very important, and that the same type of forest stand may have differing climatic influences at different elevations, within different climatic zones, and at different times of the year. The nature of this organization of the stand in space depends in turn on the vegetational species, the ecological associations, the age of the stand, and other botanical considerations. Much of the climatic influence of a forest can be explained in simple terms of the geometry of the forest, including morphological characteristics, size, coverage, and stratification. Morphological characteristics include amount of branching (bifurcation), the periodicity of growth (that is, evergreen or deciduous), together with the size, density, and texture of the leaves. Tree size is obviously important, but this may be greatly confused, particularly in tropical forests, by the great variety of sizes present locally. Crown coverage is important in terms of the physical obstruction presented by the canopy to radiation exchange and air movements. Considerations of stratification imply that one needs a particularly detailed view of the vertical forest structure in considering its climatic effects.

An obvious example of the microclimatic effects of different spatial forest organizations can be gained from a comparison of the features of tropical rain forests and temperate forests. In tropical forests the average height of the taller trees is of the order of 46–55 meters (150–180 ft), individuals rising to over 60 meters (200 ft). The dominant height of temperate forests is up to 30 meters, so that neither temperate nor most tropical forests can compare in height with the western American redwoods (*Sequoia sempervirens*). Tropical forests commonly possess a great variety of species, there being seldom less than 40 per hectare (100 hectares = 1 km²) and sometimes over 100; comparing with less than 25 (occasionally only one tree species with a trunk diameter greater than 10 cm or 4 in.) in Europe and North America. Many British woodlands, for example, have almost continuous canopy stratification from low shrubs to the tops of 36 meter beeches, whereas tropical forests are strongly stratified with dense undergrowth, unbranching trunks, and commonly two upper strata of foliage. This stratification, the second or lower of which is usually the more dense, results in rather more complex microclimates in tropical forests than in temperate stands.

It is convenient to describe the climatic effects of forest stands in terms of their modification of energy transfers and of the airflow, their modification of the humidity environment, and their modification of the thermal environment.

1. Modification of energy transfers. A major effect of a forest canopy is to blanket both incoming and outgoing radiation. The reflectivity of vegetated surfaces to incoming radiation (or albedo) has a wide range depending on the character of the vegetation and its density. For a complete covering of short, green crops the albedo is about 26%, but for temperate forests the figures range between 10 and 20% (fir 10%, young pine 14%, young oak 18%). Species of drier regions possess much higher albedos, with figures of the order of 30–38% for desert shrubs.

Besides reflecting energy, the forest canopy traps energy, and it has been calculated that for dense red beeches (*Fagus sylvatica*) 80% of the incoming radiation is intercepted by the treetops and less than 5% reaches the forest floor. The greatest trapping occurs in sunny conditions, for when the sky is overcast the incoming diffuse radiation has greater possibility of penetration laterally to the trunk space (Fig. 7.7A). Visible light, however, does not give

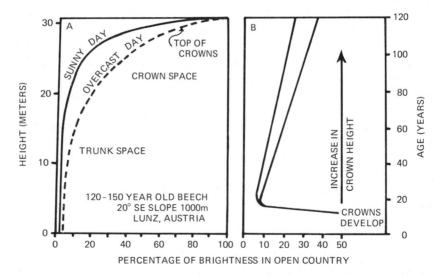

Fig. 7.7. Amount of light as a function of height (A) for a thick stand of red beeches (Fagus sylvatica) in Austria, and as a function of age (B) for a Thuringian spruce forest (after Geiger, 1965).

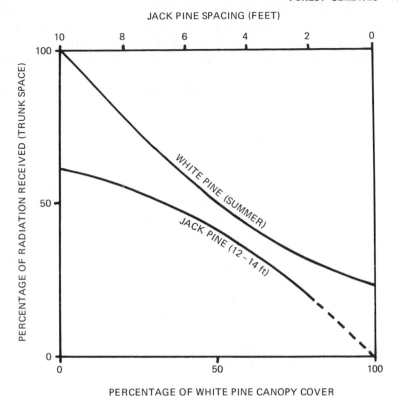

JACK PINE SPACING (FEET)

PERCENTAGE OF WHITE PINE CANOPY COVER

Fig. 7.8. Percentage of outside radiation received in pine forest trunk spaces, as a function of white pine (Pinus strobus) canopy cover in Massachusetts, and of jack pine (Pinus banksiana) spacing in Vermont (data from Kittredge, 1948).

an altogether accurate picture of total energy penetration for more ultraviolet than infrared radiation is absorbed in the crowns. For example, only 7.6% of short-wave radiation (less than 5000 Å) reached a forest floor in Nigeria, as against 45.3% greater than 6000 Å. As far as light penetration is concerned, there are obviously great variations depending on type of tree, time of year, age, crown density, and height. About 50–75% of the outside light intensity may penetrate to the floor of a birch–beech forest, 20–40% for pine, 10–25% for spruce and fir; but for tropical Congo forests the figure may be as low as 0.1% and one of 0.01% has been recorded for a dense elm stand in Germany. Of course, one of the most important effects of this is to reduce the length of daylight. For deciduous trees, more than 70% of the light may penetrate when they are leafless. Tree age is also important in that this controls both crown cover and height. Fig. 7.7B shows this rather complicated effect for spruce in the Thuringian Forest, Germany. For a Scots pine (*Pinus sylvestris*) forest in Germany 50% of the outside light intensity was recorded at 1.3 years, only 7% at 20 years, and 35% at 130 years. Fig. 7.8 shows the effects of canopy cover and tree-spacing on the amount of trunk-space radiation received. In a 25-year-old stand of red pine (*Pinus resinosa*) in Minnesota the interior light intensity fell from 60% with a tree density of 1300 trees per hectare to 15% with 6500 trees per hectare. Yet most tree-spacing is irregular and it has been estimated that only 0.5–2.5% of a forest floor in Guyana receives direct illumination.

2. Modification of the airflow. Forests impede both the lateral and vertical movement of air, but as it is more convenient to treat the latter in connection with thermal modifications we shall be concerned here with forests as obstructions to lateral air displacements. In general, air movement within forests is slight compared with that in the open, and quite large variations of outside wind velocity have little effect inside woods (Fig. 7.9A). Measurements for European forests show that 30 meters (100 ft) of penetration reduces wind velocities 60–80%, 60 meters (200 ft) to 50%, and 120 meters (400 ft) to only 7%. A speed of 2.2 m/sec (4.9 mph) outside a Brazilian evergreen forest was found to be reduced to 0.5 m/sec (1.1 mph) at about 100 meters (300 ft) within, and to be negligible at 1000 meters. In the same location external storm winds of 28 m/sec (62.7 mph) were reduced to 2 m/sec (4.2 mph) some 11 km (7 miles) deep in the forest. Where there is a complex vertical structuring of the forest wind velocities become more complex. Thus whereas in the crowns (23 meters; 75 ft) of a Panama rain forest the wind velocity was 75% of that outside, it was only 20% in the undergrowth (2 meters; 6.5 ft). Other influences include the density of the stand and the season. For dense pine stands in Idaho simultaneous recordings showed the wind velocity to be 0.6 m/sec (1.3 mph) in a cut area, 0.4 m/sec (0.9 mph) half cut, and 0.1 m/sec (0.2 mph) in the uncut stand. The effect of season on wind velocities

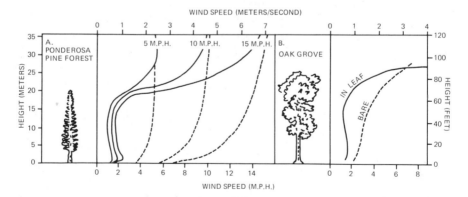

Fig. 7.9. Influence on wind velocity profiles exercised by (A) a dense 45-year-old ponderosa pine (Pinus ponderosa) stand in the Shasta Experimental Forest, California (after Fons, and Kittredge, 1948), the dashed lines indicating the wind profile over open country; and (B) an oak grove (after R. Geiger and H. Amann, and Geiger, 1965).

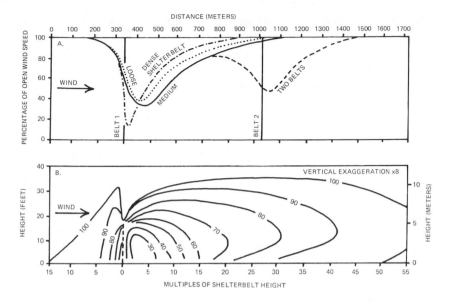

Fig. 7.10. The influence of shelterbelts on wind-velocity distributions (expressed as percentages of the velocity in the open). A. The effects of one shelterbelt of three different densities, and of two back-coupled medium-dense shelterbelts (after W. Nägeli, and Geiger, 1965). B. The detailed effects of one half-solid shelterbelt (after Bates and Stoeckeler, and Kittredge, 1948).

in deciduous forests is shown by Fig. 7.9B. Observations in a Tennessee mixed-oak forest showed January forest wind velocities to be 12% of those in the open whereas those in August had dropped to 2%.

Knowledge of the effect of forest barriers on winds has been utilized in the construction of wind breaks to protect crops and soil, and, for example, the cypress breaks of the southern Rhone valley and the Lombardy poplars (*Populus nigra*) of the Netherlands form distinctive features of the landscape. It has been found that the denser the obstruction the greater the shelter immediately behind it, although the downwind extent of its effect is reduced by lee turbulence set up by the barrier. The maximum protection is given by the filtering mechanism produced by a break of about 40% penetrability (Fig. 7.10A). An obstruction begins to have an effect about 18 times its own height upwind (Fig. 7.10B), and the downwind effect can be increased by the "back coupling" of more than one belt (Fig. 7.10A).

There are also some less obvious microclimatic effects of forest barriers. One of the most important is that the reduction of horizontal air movement in forest clearings increases the frost hazard on winter nights. A less important aspect is the removal of dust from the air by the filtering action of forests; measurements $1\frac{1}{2}$ km upwind, on the lee side, and $1\frac{1}{2}$ km downwind of a kilometer-wide German forest gave dust counts (particles per liter) of 9000, less than 2000, and more than 4000, respectively.

The catch of moisture in the form of horizontally moving fog particles is another aspect of the obstructional features of forest microclimatology. Measurements in association with a 2-meter high and 13-meter thick shelter belt on the southeast Hokkaido coast, Japan, in July 1952 showed this filtering effect on advection fog rolling in from the sea, in that 20 meters downwind of the obstruction the humidity was only 0.1 g/m³ (mean wind velocity 2.55 m/sec), compared with 0.3 g/m³ (mean wind velocity 3.4 m/sec) a similar distance upwind. In extreme cases so much fog can be filtered from laterally moving air that *negative interception* can occur, where there is a higher precipitation catch within a forest than outside. The winter rainfall catch outside a eucalyptus forest near Melbourne, Australia, was 50 cm (19.7 in.), whereas inside the forest it was 60 cm (23.6 in.).

3. Modification of the humidity environment. The humidity conditions within forest stands contrast strikingly with those in the open. Evaporation from the forest floor is usually much less because of the decreased direct sunlight, lower wind velocities, lower maximum temperatures, and generally higher forest air humidity. Evaporation from the bare floors of pine forests is 70% of that in the open for Arizona in summer and only 42% for the Mediterranean region, although such measurements have little real significance in that water losses from vegetated surfaces are controlled by the plant evapotranspiration.

During daylight leaves transpire water through open pores, or *stomata*, so that this loss is controlled by the length of day, the leaf temperature (modified by evaporational cooling), the leaf surface area, the tree species and its age, as well as by the meteorological factors of available radiant energy, atmospheric vapor pressure, and wind speed (see Chapter 2, Section A). Total evaporation figures are therefore extremely varied; also the evaporation of water intercepted by the vegetation surfaces enters into the totals, in addition to direct transpiration. Calculations made for a catchment covered with Norway spruce (*Picea abies*) in the Harz Mountains of Germany showed an estimated annual evapotranspiration of 34 cm and additional interception losses of 24 cm.

The humidity of forest stands is very much linked to the amount of evapotranspiration and increases with the density of vegetation present (Fig. 7.11A). The increase of forest humidity over that outside averages some 3–10% of relative humidity and is especially marked in summer (Fig. 7.11B). Mean annual relative humidity forest excesses for Germany and Switzerland are for beech 9.4%, Norway spruce (*Picea abies*) 8.6%, larch 7.9%, and Scots pine (*Pinus sylvestris*) 3.9%. However, humidity comparisons in these terms are rather unsatisfactory in that forest temperatures differ strikingly from those in the open. Forest vapor pressures were found to be higher within an oak stand in Tennessee than outside for every month except December. Tropical forests exhibit almost complete night saturation irrespective of

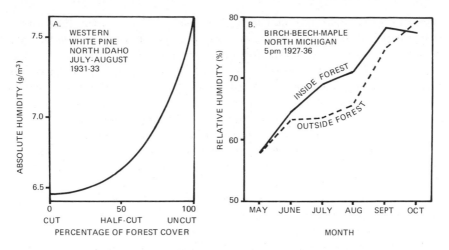

Fig. 7.11. The effects of (A) percentage of white pine (Pinus monticola) cover on summer absolute humidity in Idaho (after Kittredge, 1948); and (B) season on relative humidity in a Michigan birch–beech–maple forest (after U.S. Dept. Agriculture Yearbook 1941).

elevation in the trunk space, whereas during the day humidity is inversely related to elevation.

The influence of forest structures on precipitation is still very much an unresolved problem. This is partly due to the difficulties of comparing rain-gauge catches in the open with those near forests, within clearings, or beneath trees. For example, on the windward side of a forest the dominance of low-level upcurrents decreases the amount of precipitation actually caught in the rain gauge, whereas the reverse occurs where there are lee side down-draughts. In small clearings the low wind velocities cause little turbulence around the opening of the gauge and catches are generally greater than outside the forest, although the actual precipitation amounts may be identical. On the other hand, it is sometimes found that the larger the clearing the more prevalent are downdraughts and consequently the precipitation catch increases. In a 25-meter-high pine and beech forest in Germany, catches in clearings of 12 meters diameter were only 87% of that upwind of the forest, but this catch rose to 105% in clearings of 38 meters diameter. An analysis of precipitation records for Letzlinger Heath (Germany) before and after afforestation suggested a mean annual increase of 6%, with the greatest excesses occurring during drier years. It is generally agreed, however, that forests have little effect on cyclonic rain, but that they may have a marginal orographic effect in increasing lifting and turbulence, which is of the order of 1–3% in temperate regions.

A far more important obstructional influence of forests on precipitation is in terms of the direct interception of rainfall by the canopy. This obviously varies with crown coverage, with season, and with the rainfall intensity. Measurements in German beech forests indicate that, on average, they intercept 43% of precipitation in summer and 23% in winter. Pine forests may intercept up to 94% of low-intensity precipitation but as little as 15% of high intensities, the average for temperate pines being about 30%. The intercepted precipitation either evaporates on the canopy, runs down the trunk, or drops to the ground. Assessment of the total precipitation reaching the ground (the *throughfall*) requires very detailed measurements of the stem flow and the contribution of drips from the canopy. Canopy evaporation is not necessarily a total loss of moisture from the woodland, since the solar energy used in the evaporating process is not available to remove soil moisture or transpiration water, but the vegetation does not derive the benefit of the cycling of the water through it via the soil. Evaporation from the canopy is very much a function of net radiation receipts (20% of the total precipitation evaporates from the canopy of Brazilian evergreen forests), and of the type of species. Some Mediterranean oak forests yield virtually no stemflow and their 35% interception almost all evaporates from the canopy. Recent investigations of the water balance of forests provide some evidence that evergreen forests may be subject to greater evapotranspiration than grass in the same climatic conditions. Grass normally reflects 10–15% more solar radiation than coniferous tree species and hence less energy is available for evaporation. In addition evergreens allow transpiration to occur throughout the year. Nevertheless many more detailed and careful studies are required to check these results and to test the various hypotheses.

4. Modification of the thermal environment. From what has been said it is apparent that forest vegetation has an important effect on microscale temperature conditions. Shelter from the sun, blanketing at night, heat loss by evapotranspiration, reduction of wind speed, and the impeding of vertical air movement all influence the temperature environment. The most obvious effect of canopy blanketing is that inside the forest daily maximum temperatures are lower and minima are higher (Fig. 7.12A). This is particularly apparent during periods of high summer evapotranspiration which depress daily maximum temperatures and cause mean monthly temperatures in tropical and temperate forests to fall well below those outside. In temperate forests at sea level the mean annual temperature may be about 0.6°C (1°F) less than that in surrounding open country, the mean monthly differences may reach 2.2°C (4°F) in summer but not exceed 0.1°C in winter, and on hot summer days the difference can be more than 2.8°C (5°F). Mean monthly temperatures and temperature ranges for temperate beech, spruce, and pine forests are given in Fig. 7.12B and C, which also show that when trees do not transpire greatly in the summer (for example, the *Forteto* oak maquis of the Mediterranean) the high day temperatures reached in the sheltered woods may cause mean monthly figures to reverse the trend exhibited by temperate forests. Even within individual climatic regions it is difficult to generalize, however, for at elevations of 100 meters the lowering of temperate forest

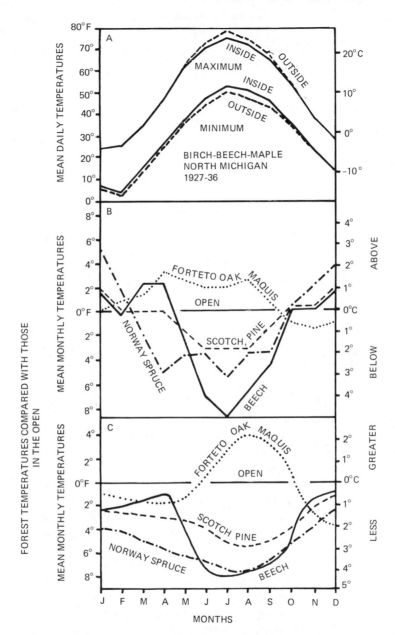

Fig. 7.12. Seasonal regimes of forest temperatures. A. Mean daily maximum and minimum temperatures inside and outside a birch–beech–maple forest in Michigan (after U.S. Dept. Agriculture Yearbook 1941). B. Mean monthly temperatures and C. mean monthly temperature ranges, compared with those in the open, for four types of Italian forest (FAO, 1962). Note the anomalous conditions associated with the forteto oak maquis, which transpires little.

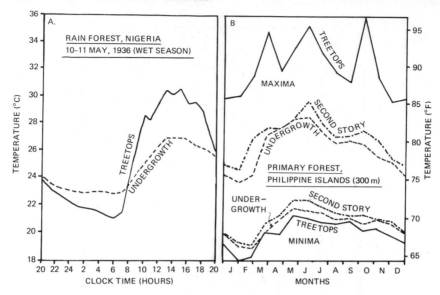

Fig. 7.13. *The effect of tropical rain forest stratification on temperature (after Richards, 1952). A. Daily march of temperature (10–11 May 1936) in the treetops (24 meters) and in the undergrowth (0.7 meters) during the wet season in primary rain forest at Shasha Reserve, Nigeria (after Evans). B. Average weekly maximum and minimum temperatures in three layers of primary (Dipterocarp) forest, Mt Maquiling, Philippine Islands (after Brown).*

mean temperatures below those in open country may be double that at sea level.

The complex vertical structure of forest stands is a further factor in complicating forest temperatures. Even in relatively simple stands vertical differences are very apparent. For example, in a ponderosa pine forest (*Pinus ponderosa*) in Arizona the recorded mean June–July maximum was increased by 0.8°C (1.5°F) simply by raising the thermometer from 1.5 to 2.4 meters (5 to 8 ft) above the forest floor. In stratified tropical forests the thermal picture is much more complicated. The dense canopy heats up very much during the day and quickly loses its heat at night, showing a much greater diurnal temperature range than the undergrowth (Fig. 7.13A). Whereas daily maximum temperatures of the second story are intermediate between those of the treetops and the undergrowth, the nocturnal minima are higher than either treetops or undergrowth because the second story is insulated by trapped air both above and below (Fig. 7.13B).

8

Climate Variability, Trends, and Fluctuations

Probably the aspect of climate which most interests the layman is speculation regarding its possible trends. Unfortunately, as well as being the most interesting, it is also the most uncertain aspect of meteorological research. Realization that climate has changed radically with time came only during the 1840s when indisputable evidence of former ice ages was obtained, yet in many parts of the world the climate has altered sufficiently, even within the last few thousand years, to affect the possibilities for agriculture and settlement. Reliable weather records have only been kept during the last hundred years or so and therefore it is only the recent climatic fluctuations which can be investigated adequately. The discussion in this chapter is mainly limited to these events, but first it is worthwhile considering the methods of handling the meteorological records which are available.

A. CLIMATIC DATA

1. Averages. The climate of a place is often regarded simply as its "average weather," but vital climatic information is overlooked if the range and frequency of extremes are neglected. Averages can be markedly affected by extreme values and this is particularly true of the arithmetic mean (which is obtained by totaling the individual values and dividing by the number of occurrences). For this reason a thirty or thirty-five-year period is normally required for the determination of climatic averages. Even so, certain types of data are very inadequately summarized by the arithmetic mean, especially when small values are frequent but very large ones occur occasionally. This situation is illustrated by Fig. 8.1, where the histogram or frequency-distribution graph of annual rainfall at Muscat (Arabia) is clearly dissimilar to that for Padua (Italy). The latter approximates a symmetrical normal

271

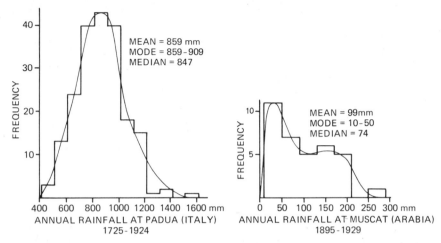

Fig. 8.1. Frequency histograms of annual rainfall in Padua, Italy (1725–1924) and Muscat, Arabia (1895–1929) (based on data from Conrad and Pollak, 1950 and 'World Weather Records'). The smooth curves indicate the approximate frequency curve at each station.

distribution, where half the values lie above and half below the mean and where the most frequent (or *modal*) category is equal to the mean.[1] The graph for Muscat, on the other hand, exhibits *positive skewness,* or the tail of the distribution is towards values greater than the mean.

One other useful measure of central tendency is the median value, which has exactly half the number of items in the data series above it and half below it. Consequently it is not biased by the occurrence of a few extreme figures. At Muscat in Fig. 8.1B, for example, the mean is a third greater than the median.

2. Variability. Variability about the average can be expressed in several ways. When the median is used it is common also to determine the upper and lower quartiles (Q_1 and Q_2), which are the central values between the median and the upper and lower extremes, respectively. The average deviation from the median is given by $(Q_1 - Q_2)/2$. A more widely used measure of variability is the standard deviation (σ, pronounced *sigma*) which is calculated by summing the square of the deviation of each value from the mean, dividing by the number of cases and then taking the square root.

$$\sigma = \sqrt{\left\{\frac{\Sigma\,(x_i - \bar{x})^2}{n}\right\}},$$

[1] Fuller details of elementary statistical procedures may be found in S. Gregory, *Statistical Methods and the Geographer* (Longmans), 1963, or M. J. Moroney, *Facts from Figures* (Pelican), 1951 and later editions.

where

$$x_i = \text{an individual value}$$

$$\bar{x} = \frac{\Sigma \, x_i}{n} \text{ (the mean)}$$

$$n = \text{number of cases}$$

$$\Sigma = \text{sum of all values.}$$

It is therefore a measure of average deviation, where the difficulty created by positive and negative departures (that is, values greater and less than the mean) is removed by squaring each deviation and rectifying this by finally calculating the square root.

Variability of rainfall may be compared between stations if the standard deviation is expressed as a percentage of the mean; the *coefficient of variation*,

$$CV = \frac{\sigma}{\bar{x}} \times 100(\%).$$

Where the standard deviation is not available, use has often been made in the past of the mean deviation, M.D.

$$MD = \frac{\Sigma |x_i - \bar{x}|}{n}$$

where $|-|$ denotes the absolute difference, disregarding sign. This measure of variability, standardized against the mean, ranges for annual precipitation from about 10–20% in western Europe and parts of monsoon India to over 50% in arid areas of the world (Fig. 8.2). It is in such areas that a small

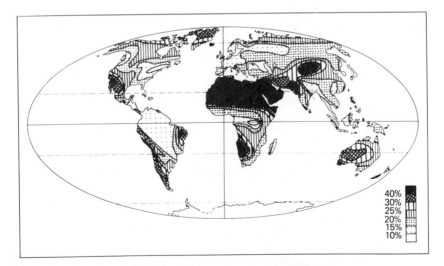

40%	
30%	
25%	
20%	
15%	
10%	

Fig. 8.2. Distribution of rainfall variability over the continents (after Erwin Biel).

change in the frequency of rainstorms can markedly affect the "average" rainfall over a given period of years. It should be noted, however, that detailed examination of the precipitation in many diverse climatic regions show that the apparent inverse relationship between annual total and variability is only very approximate. Moreover, a coefficient of variation $\geq 50\%$ in fact violates the statistical assumption of a normal frequency distribution on which this statistic is based.

3. Trends. It is obvious that the great year-to-year variability of climatic conditions may conceal gradual trends from one type of regime towards another. The effect of short-term irregularities can be removed by various statistical techniques, of which the simplest is the *running mean* (or *moving average*). The method is to calculate mean values for successive, overlapping periods of perhaps five, ten, or thirty years, for example,

$$\frac{\text{Year } 1 + \text{Year } 2 + \text{Year } 3 + \text{Year } 4 + \text{Year } 5}{5} = \text{Mean for Year 3}$$

$$\frac{\text{Year } 2 + \text{Year } 3 + \text{Year } 4 + \text{Year } 5 + \text{Year } 6}{5} = \text{Mean for Year 4,}$$
$$\text{and so on.}$$

This device smooths out the short-term fluctuations if periods of twenty or thirty years are used, thereby emphasizing the long-term trends. Callendar has used this method to show that there has been an increase in warmth since the end of the last century, particularly in the north temperate zone

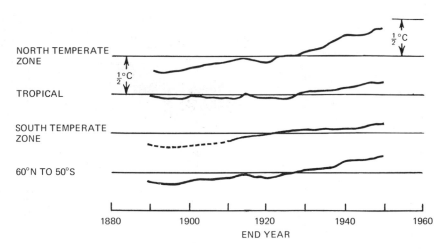

Fig. 8.3. *Temperature trends for generalized latitude zones. The graphs show the departure of 20-year moving averages from the mean temperature for 1901–30 (from Callendar, 1961).*

(Fig. 8.3). The change appears to have been least in the tropics and the largest in the Sub-Arctic; the changes of annual mean temperature between the thirty-year periods 1890–1919 and 1920–49 were approximately +0.2°C (0.3°F) for the tropics and +0.4°C (0.6°F) for the north temperate zone, while the rise in the Svalbard–Greenland Sea area amounted to +2.8°C (5.0°F) between 1900–19 and 1920–39.

These changes represent an increased frequency of warm months rather than a slight overall temperature increase. For instance, the records at Copenhagen (Table 8.1) show that warm winter and summer months have become more frequent, and cold ones less frequent, during the present century.

Table 8.1:   The frequency of January and July temperatures at Copenhagen (*after Lysgaard, 1963*).

January Mean (°C)	1811–40	1841–70	1871–1900	1901–30	1931–60
≤ −1.5	14	13	6	6	4
−1.4 to 1.4	11	12	15	17	18
≥ 1.5	5	5	7	7	8
July Mean (°C)					
≤ 15.5	7	9	4	3	1
15.6 to 18.4	18	19	24	22	21
≥ 18.5	5	2	2	5	8

B. THE CLIMATIC RECORD

To understand the significance of climatic trends over the last hundred years they need to be viewed against the background of our general knowledge about the postglacial climatic record. The available information is based on archaeological evidence, analysis of the vegetation history by means of pollen remains preserved in peat bogs, and, for more recent periods, documentary sources. Although these rarely provide precise quantitative data the general pattern is nevertheless clear.

1. The major postglacial epochs. Following the final retreat of the continental ice sheets from Europe and North America between 10,000 and 7000 years ago the climate rapidly ameliorated in middle and higher latitudes. A thermal maximum was reached about 5000 to 3000 B.C. when summer temperatures are known to have been several degrees higher than today. Thereafter a decline set in with cold, wet conditions in Europe around 900–500 B.C. Although temperatures have not since equaled those of the thermal maximum there was certainly a warmer period in many

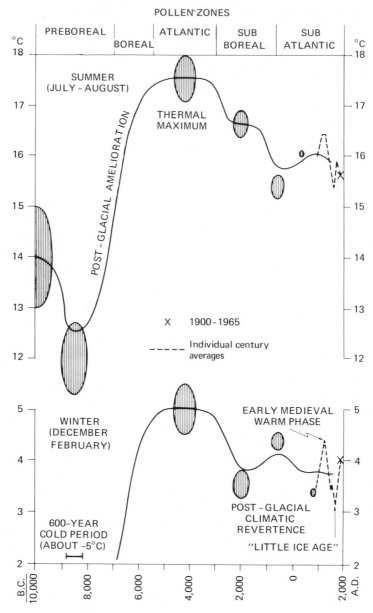

POLLEN ZONES

Fig. 8.4. Air temperatures in the lowlands of central England. Trends of the supposed 1000-year and 100-year averages since 10,000 B.C. (the latter calculated for the last millennium) (after Lamb, 1966). Shaded ovals indicate the approximate ranges within which the temperature estimates lie and error margins of the radiocarbon dates. Note that the preboreal phase begins about 8300 B.C. following the end of the Glacial Period.

parts of the world between about A.D. 1000 and 1250, and this phase was marked by the Viking colonization of Greenland and the occupation of Ellesmere Island in the Canadian Arctic by Eskimos. A further deterioration followed and severe winters between A.D. 1550 and 1700 gave a Little Ice Age with extensive Arctic pack-ice and glacier advances in some areas to maximum positions since the end of the Ice Age. These advances occurred at dates ranging from the mid-seventeenth to the late nineteenth century in different areas, as a result of the lag in glacier response and minor climatic fluctuations. Fig. 8.4 attempts to summarize these trends, but it must be stressed that at present only the gross features are represented, as we know little or nothing about short-term fluctuations before the Medieval period, for example, and even the relative magnitudes of the changes prior to about A.D. 1700 can only be indicated in a very general way.

2. The recent warming trend. Long instrumental records for a few stations in Europe indicate that the warming trend which ended the Little Ice Age began early in the nineteenth century, although it was interrupted in some areas about 1880–90 (Fig. 8.5). Winter temperatures have been most affected and on Svalbard the twenty-year change in January mean temperature from 1900–19 to 1920–39 amounted to +7.8°C (14.0°F).

The effects of the temperature rise are apparent in many ways. There has been, for example, a rapid retreat of most of the world's glaciers (for example, measurement of the Muir Glacier in Alaska shows it has retreated 3 km in 10 years). Glaciers in the North Atlantic area seem to be universally shrinking at present, due largely to temperature increases which have the effect of lengthening the ablation season, rather than to decreases in snowfall. The world rise in temperature has caused a corresponding raising of the snowline. Since 1900 it has risen 800 meters in Peru, and in Sweden the timberline on some mountain slopes has risen as much as 20 meters since 1930. Associated with the melting of glaciers has been a general rise in sea level which has been estimated at about 15 cm between 1930 and 1948, a rise of four times the average rate experienced during the last 9000 years. The rate became six times the average in the mid-1920s, the time when glacier melting showed a particularly marked increase. Another tendency illustrating world warming has been the retreat of the tundra margin. In Alaska and Siberia trees are beginning to annex former tundra areas to the north. In Siberia the southern limit of permanently frozen ground is retreating northwards at the rate of several dozen meters a year. Ports in the Arctic are remaining free of ice for longer periods each year and cod have extended their feeding northwards off west Greenland by 9° of latitude between 1919 and 1948.

Unfortunately the latest evidence suggests that the warm period of the 1920s and '30s has come to an end. Cooling has taken place especially over northern Siberia, the eastern Canadian Arctic, and Alaska, with changes of the order of −2° to −3°C (−4° to −6°F) in winter mean temperature

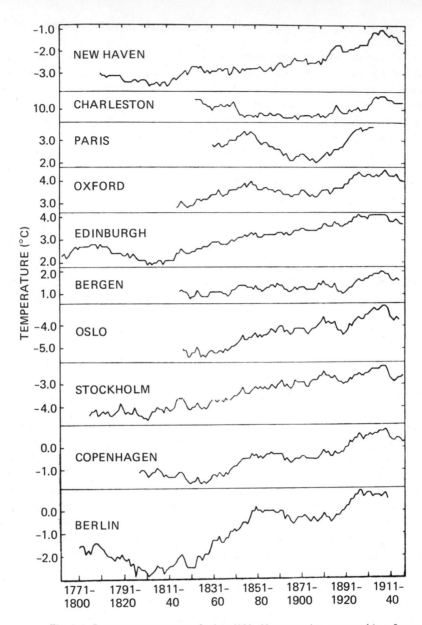

Fig. 8.5. *January temperature trends since 1800. 30-year moving averages (from Lysgaard, 1949).*

from 1940–9 to 1950–9. Perhaps in compensation slight winter warming has at the same time affected the United States, eastern Europe, and Japan. Whether or not this "downturn" of temperature represents a minor fluctuation or a longer-term trend remains to be seen, but it is clear that the latter would have important economic implications in many parts of the northern hemisphere.

3. Recent changes in tropical rainfall. While temperatures in the tropics seem to have undergone little change since the 1880s, the same is not true of rainfall totals. Over extensive areas there is evidence of a marked decrease of annual rainfall at the end of the nineteenth century. Kraus illustrates the magnitude of this change at Freetown (Sierra Leone):

Seasonal rainfall averages at Freetown.

	May–October	*November–April*
1875–1896	452.3 cm (178.05 in.)	41.7 cm (16.41 in.)
1907–1931	312.1 cm (122.85 in.)	23.1 cm (9.10 in.)

The early rains (May–June) and the late rains (September–October) show a greater *relative* decrease than the very wet months of July and August, which points to a lengthening of the dry season as the main factor.

The greater aridity in the tropics was apparently not accompanied by any compensating increase in temperate latitudes, indeed rainfall also decreased in southeast Australia and eastern North America. This suggests that a general weakening of the moisture cycle, particularly a reduction of evaporation, was probably responsible, but the underlying reason for such a change is not yet clear.

C. POSSIBLE CAUSES OF CLIMATIC CHANGE

The immediate cause of the recent climatic fluctuations appears to be the strength of the global wind circulation. The first thirty years of this century saw a pronounced increase in the vigor of the westerlies over the North Atlantic, the northeast trades, the summer monsoon of southern Asia, and the southern hemisphere westerlies (in summer). In the northeast North Atlantic this trend was already under way early in the nineteenth century. It has been accompanied by more northerly depression tracks across the Atlantic and by independent changes in the meridional (north–south) components of airflow. A comparison of pressure readings for the period 1900–19 with the period 1920–39 shows that the Siberian high-pressure cell spread westward (5–10° of longitude) towards Scandinavia. At the same time the Icelandic and winter Mediterranean low-pressure cells deepened (decreasing by 2 mb and 1 mb, respectively) and the North American high-pressure cell may have spread eastward. The Aleutian and Icelandic low-pressure cells also tended to move 5–10° of longitude eastward.

The picture over the North Atlantic thus consisted of an increased pressure gradient between the Azores high and the Icelandic low, on the one hand, and between the Siberian high and the Icelandic low on the other. This resulted in a significant increase in the frequency of mild southwesterly airflow over the British Isles between about 1900 and 1930, as reflected by the average annual frequency of Lamb's Westerly airflow type (see Chapter 5, Section A.3). For 1873–97, 1898–1937, and 1938–61, the figures are 27%, 38%, and 30%, respectively. The recent decrease of westerly airflow, especially in winter, is linked with an increase of northerly airflow, giving more frequent snowfalls, while the southward shift of the main depression tracks has produced a number of cool, wet summers in Britain (notably 1954, 1956, 1958, and 1960).

The key to these atmospheric variations must be linked to the heat balance of the earth-atmosphere system and this forces us to return to the fundamental energy considerations with which we began this book. It is possible that there have been changes in the amount of energy sent out by the sun, but the present evidence for fluctuations in the solar constant is inconclusive. However, variations apparently do occur in the emission of high-energy particles and ultraviolet radiation during brief solar flares. All solar activity follows the well-known cycle of approximately eleven years, which is usually measured with reference to the period between sunspot maximum and minimum and, although numerous attempts have been made to correlate sunspot numbers with meteorological events, the results so far obtained are conflicting. For instance, tropical temperatures show positive correspondence with sunspot numbers between 1930 to 1950, but from 1875 to 1920 the relationship was an inverse one. Such phase changes may be capable of interpretation when more is known about the effect of solar activity on the upper atmosphere, although only very tentative hypotheses have yet been advanced. A recent one suggests that stratospheric ozone is more abundant about two years before a sunspot minimum. The latter is accompanied by lower solar latitudes of sunspot activity causing solar radiation to be pointed more directly at the earth. The resulting increase of ultraviolet radiation yields more ozone with consequently greater absorption of solar insolation. It is postulated that warming of the stratosphere weakens the subtropical high-pressure belt and in turn the westerlies, giving lower than average rainfall in temperate oceanic climates. In continental areas the weaker westerlies may favor greater convectional rainfall, so that complex phase relationships can be envisaged between weather and the sunspot cycle.

Another possible factor, which is not necessarily an alternative, is a change in the tropospheric heat budget. It has previously been noted that the amount of carbon dioxide in the atmosphere seems to bear a direct relationship with temperature. In the last forty years world measurements of carbon dioxide have shown an increase of 10% (see Fig. 1.3). This would naturally lead to an increase of heat absorption by the atmosphere from the earth's outgoing long-wave reradiation. Yet it is questionable whether this result is based on readings which are truly representative of the whole world. Most of them are necessarily restricted to the inhabited areas of the northern hemisphere. One strong objection to this is that the polar areas (where the greatest temperature changes have been recorded) are notably deficient in carbon dioxide. Moreover, as pointed out earlier, the warming trend of the early part of this century now appears to have ended!

Other factors may also have modified the atmospheric heat budget. The presence of increased amounts of volcanic dust in the atmosphere is one suggested cause of the Little Ice Age and, while this may be the case, it seems doubtful that reduced dust content was responsible for the early twentieth-century warming, since the trend was least in the tropics and greatest in cloudy, maritime regions of high latitudes. In fact this pattern is strong evidence of a heat-transporting mechanism. Increased storminess might cause more frequent movement of warm air masses towards high latitudes, yet increases in winter temperatures also occur in areas which are predominantly affected by northerly winds and therefore cannot receive heat directly from the source regions of warm air.

The interaction of ocean and atmosphere may be responsible for some of these complications. The sea can store vast quantities of heat and may therefore greatly modify the heat and moisture exchanges with the air above. However, recent detailed investigations by J. Bjerknes show that variations in Atlantic sea-surface temperatures are due to initial changes in the wind regime. Hence, although feedback effects may operate, the atmospheric changes appear to be the primary mechanism. In the short-term these may be the result of inherent instability in the atmospheric circulation, but the occurrence of Ice Ages probably demands the existence of distinct causative factors. Whether these are terrestrial or extraterrestrial remains a problem for future research to solve.

Bibliography

General

BARRETT, E. C. 1967, *Viewing Weather from Space;* (Longmans, London), 140 pp.

BATES, D. R. 1957, *The Planet Earth;* (Pergamon Press, London), 312 pp.

* BATTAN, L. J. 1959, *Radar Meteorology;* (Univ. of Chicago Press, Chicago), 161 pp.

* BERRY, F. A., BOLLAY, E., and BEERS, N. R. (*Eds.*) 1945, *Handbook of Meteorology;* (McGraw-Hill, New York), 1068 pp.

BLAIR, T. A. and FITE, R. C. 1965, *Weather Elements; A text in elementary meteorology;* 5th ed. (Prentice-Hall, New York), 364 pp.

BLÜTHGEN, J. 1966, *Allgemeine Klimageographie;* 2d ed. (W. de Gruyter, Berlin), 720 pp.

BRUCE, J. P. and CLARK, R. H. 1966, *Introduction to Hydrometeorology;* (Pergamon Press, London), 319 pp.

* BYERS, H. R. 1959, *General Meteorology;* 3d ed. (McGraw-Hill, New York), 540 pp.

* DOBSON, G. M. B. 1963, *Exploring the Atmosphere;* (Oxford University Press), 188 pp.

DONN, W. T. 1965, *Meteorology;* 3d ed. (McGraw-Hill, New York), 484 pp.

ENVIRONMENTAL SCIENCE SERVICES ADMINISTRATION 1965, *APT Users' Guide;* U.S. Dept. of Commerce, (Washington D.C.), 100 pp.

FAIRBRIDGE, R. W. (*Ed.*) 1967, *The Encyclopedia of Atmospheric Sciences and Astrogeology;* (Rheinold, New York), 1200 pp.

FLOHN, H. 1968, *Climate and Weather;* (World University Press), 253 pp.

FRITZ, S. 1964, Pictures from meteorological satellites and their interpretation; *Space Science Reviews*, Vol. 3, p. 541–580.

GEIGER, R. 1965, *The Climate near the Ground;* 2d ed. (Harvard), 611 pp.

HALTINER, G. J. and MARTIN, F. L. 1957, *Dynamical and Physical Meteorology;* (McGraw-Hill, New York), 470 pp.

HARE, F. K. 1953, *The Restless Atmosphere;* (Hutchinson, London), 192 pp.

HARE, F. K. 1955, Dynamic and synoptic climatology; *Ann. Assn. Amer. Geog.*, Vol. 45, p. 152–162.

---

* Indicates more advanced texts with reference particularly to the physical bases of meterology, often including mathematical material.

HARE, F. K. 1957, The dynamic aspects of climatology; *Geografiska Annaler*, Vol. 39, p. 87–104.

HAURWITZ, B. and AUSTIN, J. M. 1944, *Climatology;* (McGraw-Hill, New York), 410 pp.

* HESS, S. L. 1959, *Introduction to Theoretical Meteorology;* (Henry Holt, New York), 362 pp.

HEWSON, E. W. and LONGLEY, R. W. 1944, *Meteorology, Theoretical and Applied;* (Wiley and Sons, New York), 468 pp.

* HUMPHREYS, W. J. 1929, *Physics of the Air;* (McGraw-Hill, New York), 654 pp.

HUSCHKE, R. E. (*Ed.*) 1959, *Glossary of Meteorology;* (American Meteorological Society, Boston), 638 pp.

HUTCHINSON, G. E. 1957, *A Treatise on Limnology;* Vol. 1 (Wiley, New York), 1015 pp.

KENDREW, W. G. 1949, *Climatology;* (Oxford), 383 pp. (2d ed., 1957, 400 pp.).

KENDREW, W. G. 1953, *The Climates of the Continents;* 4th ed. (Oxford), 607 pp.

LAMB, H. H. 1964, *The English Climate;* (English Universities Press), 212 pp.

LIST, R. J. 1951, *Smithsonian Meteorological Tables;* 6th ed. (Smithsonian Institution, Washington), 527 pp.

MCINTOSH, D. H. 1963, *Meteorological Glossary;* 2d ed. (H.M.S.O., London), 288 pp.

* MALONE, T. F. (*Ed.*) 1951, *Compendium of Meteorology;* (American Meteorological Society, Boston), 1334 pp.

MANLEY, G. 1952, *Climate and the British Scene;* (Collins, London), 314 pp.

METEOROLOGICAL OFFICE 1958, *Tables of Temperature, Relative Humidity, and Precipitation for the World;* 6 vols. (H.M.S.O., London).

MILLER, A. 1966, *Meteorology;* (Merrill, Columbus, Ohio), 128 pp.

MILLER, A. A. 1957, *Climatology;* 8th ed. (Methuen, London), 318 pp.

MILLER, A. A. and PARRY, M. 1958, *Everyday Meteorology;* (Hutchinson, London), 270 pp.

MUNN, R. E. 1966, *Descriptive Micrometeorology;* (Academic Press, New York), 245 pp.

NATIONAL AERONAUTICS AND SPACE ADMINISTRATION 1966, *NIMBUS II Users' Guide;* (Goddard Space Flight Center, Greenbelt, Maryland), 229 pp.

PEDGLEY, D. E. 1962, *A Course of Elementary Meteorology;* (H.M.S.O., London), 189 pp.

* PETTERSSEN, S. 1956, *Weather Analysis and Forecasting;* 2 vols. (McGraw-Hill, New York), 428 and 266 pp.

PETTERSSEN, S. 1958, *Introduction to Meteorology;* 2d ed. (McGraw-Hill, New York), 327 pp.

* REITER, E. R. 1963, *Jet Stream Meteorology;* (University of Chicago Press), 515 pp.

RIEHL, H. 1965, *Introduction to the Atmosphere;* (McGraw-Hill, New York), 365 pp.

* SELLERS, W. D. 1965, *Physical Climatology;* (University of Chicago Press), 272 pp.

STRAHLER, A. N. 1965, *Introduction to Physical Geography;* (Wiley, New York), 455 pp.

SUTCLIFFE, R. C. 1941, *Meteorology for Aviators;* (H.M.S.O., London), 274 pp. (and later editions).

SUTTON, O. G. 1960, *Understanding Weather;* (Pelican, London), 215 pp.

SUTTON, O. G. 1962, *The Challenge of the Atmosphere;* (Hutchinson, London), 227 pp.

SVERDRUP, H. V. 1945, *Oceanography for Meteorologists;* (Allen & Unwin, London), 235 pp.

SVERDRUP, H. V., JOHNSON, M. W., and FLEMING, R. H. 1942, *The Oceans; Their physics, chemistry and general biology;* (Prentice-Hall, New York), 1087 pp.

TAYLOR, J. A. and YATES, R. A. 1967, *British Weather in Maps;* 2d ed. (Macmillan, London), 315 pp.

TREWARTHA, G. T. 1954, *An Introduction to Climate;* (McGraw-Hill, New York), 402 pp.

TREWARTHA, G. T. 1961, *The Earth's Problem Climates;* (McGraw-Hill, New York), 334 pp.

VAN ROOY, M. P. 1957, *Meteorology of the Antarctic;* (Weather Bureau, Pretoria), 240 pp.

WILLETT, H. C. and SANDERS, F. 1959, *Descriptive Meteorology;* 2d ed. (Academic Press, New York), 355 pp.

WORLD METEOROLOGICAL ORGANIZATION 1962, *Climatological Normals (CLINO) for CLIMAT and CLIMAT SHIP stations for the period 1931–60;* (World Meteorological Organization, Geneva).

Chapter 1. Atmospheric energy

BARRY, R. G. and CHAMBERS, R. E. 1966, A preliminary map of summer albedo over England and Wales; *Quart. J. Roy. Met. Soc.*, Vol. 92, p. 543–548.

BECKINSALE, R. P. 1945, The altitude of the zenithal sun: A geographical approach; *Geog. Rev.*, Vol. 35, p. 596–600.

BUDYKO, M. I. *et al.*, 1962, The heat balance of the surface of the earth; *Soviet Geography*, Vol. 3(5), p. 3–16.

CRAIG, R. A. 1965, *The Upper Atmosphere; Meteorology and Physics;* (Academic Press, New York), 509 pp.

GARNETT, A. 1937, Insolation and relief; *Trans. Inst. Brit. Geog.*, No. 5, 71 pp.

GODSON, W. L. 1960, Total ozone and the middle stratosphere over arctic and subarctic areas in winter and spring; *Quart. J. Roy. Met. Soc.*, Vol. 86, p. 301–317.

HARE, F. K. 1962, The stratosphere; *Geog. Rev.*, Vol. 52, p. 525–547.

HASTENRATH, S. L. 1968, Der regionale und jahrzeitliche Wandel des vertikalen Temperaturgradienten und seine Behandlung als Wärmhaushaltsproblem; *Meteorologische Rundschau*, Vol. 21, p. 46–51.

KUNG, E. C., BRYSON, R. A., and LENSCHOW, D. H. 1964, Study of a continental surface albedo on the basis of flight measurements and structure of the earth's surface cover over North America; *Monthly Weather Review*, Vol. 92, p. 543–564.

LAUTENSACH, H. and BOGEL, R. 1956, Der Jahrsgang des mittleren geographischen Hohengradienten der Lufttemperatur in den verschiedenen Klimagebieten der Erde; *Erdkunde*, Vol. 10, 270–282.

LUMB, F. E. 1961, *Seasonal variation of the sea surface temperature in coastal waters of the British Isles;* Sci. Paper No. 6, Meteorological Office (H.M.S.O., London), 21 pp.

MCFADDEN, J. D. and RAGOTZKIE, R. A. 1967, Climatological significance of albedo in central Canada; *J. Geophys. Res.*, Vol. 72, p. 1135–1143.

MILLER, D. H. 1968, A Survey Course: The Energy and Mass Budget at the Surface of the Earth; *Pub. No. 7, Amer. Assn. Geog.* (Washington, D.C.), 142 pp.

NEWELL, R. E. 1963, Transfer through the tropopause and within the stratosphere; *Quart. J. Roy. Met. Soc.*, Vol. 89, p. 167–204.

NEWELL, R. E. 1964, The circulation of the upper atmosphere; *Sci. American*, Vol. 210, p. 62–74.

NEWTON, H. W. 1958, *The Face of the Sun;* (Pelican, London), 208 pp.

PAFFEN, K. 1967, Das Verhältniss der Tages- zur Jahrzeitlichen Temperaturschwankung; *Erdkunde*, Vol. 21, p. 94–111.

PLASS, G. M. 1959, Carbon Dioxide and Climate; *Sci. American*, Vol. 201, p. 41–47.

RADOS, R. M. 1967, The evolution of the TIROS meteorological satellite operational system; *Bull. Am. Met. Soc.*, Vol. 48, p. 326–337.

RANSOM, W. H. 1963, Solar radiation and temperature; *Weather*, Vol. 8, p. 18–23.

RATCLIFFE, J. A. (*Ed.*) 1960, *Physics of the Upper Atmosphere;* (Academic Press, New York & London), 586 pp.

SHEPPARD, P. A. 1968, Global atmospheric research; *Weather*, Vol. 23, p. 262–283.

STONE, R. 1955, Solar heating of land and sea; *Geography*, Vol. 40, p. 288.

SUTTON, G. 1963, Scales of temperature; *Weather*, Vol. 18, p. 130–134.

VALLEY, S. L. (*Ed.*) 1965, *Handbook of Geophysics and Space Environments;* (McGraw-Hill, New York).

WORLD METEOROLOGICAL ORGANIZATION, 1964, Regional basic networks; *W.M.O. Bulletin*, Vol. 13, p. 146–147.

Chapter 2. Atmospheric moisture

ARMSTRONG, C. F. and STIDD, C. K. 1967, A moisture-balance profile in the Sierra Nevada; *J. of Hydrology*, Vol. 5, p. 258–268.

BERGERON, T. 1960, Problems and methods of rainfall investigation; In *The Physics of Precipitation*, Geophysical Monograph No. 5, Amer. Geophys. Union (Washington), p. 5–30.

BRAHAM, R. R. 1959, How does a raindrop grow?; *Science*, Vol. 129, p. 123–129.

BYERS, H. R. and BRAHAM, R. R. 1949, *The Thunderstorm;* U.S. Weather Bureau.

DURBIN, W. G. 1961, An introduction to cloud physics; *Weather*, Vol. 16, p. 71–82 and 113–125.

EAST, T. W. R. and MARSHALL, J. S. 1954, Turbulence in clouds as a factor in precipitation; *Quart. J. Roy. Met. Soc.*, Vol. 80, p. 26–47.

GARCIA-PRIETO, P. R., LUDLAM, F. H., and SAUNDERS, P. M. 1960, The possibility of artificially increasing rainfall on Tenerife in the Canary Islands; *Weather*, Vol. 15, p. 39–51.

GILMAN, C. S. 1964, Rainfall; In *Handbook of Applied Hydrology* (*Ed.* Chow, V. T.) (McGraw-Hill, New York), Section 9.

HARROLD, T. W. 1966, The measurement of rainfall using radar; *Weather*, Vol. 21, p. 247–249 and 256–258.

HASTENRATH, S. L. 1967, Rainfall distribution and regime in Central America; *Archiv für Meteorologie, Geophysik und Bioklimatologie*, Series B, Vol. 15(3), p. 201–241.

HERSHFIELD, D. M. 1961, Rainfall frequency atlas of the United States for durations from 30 minutes to 24 hours and return periods of 1 to 100 years; *U.S. Weather Bureau, Tech. Rept.* 40.

HOPKINS, M. M. JR. 1967, An approach to the classification of meteorological satellite data; *J. Appl. Met.*, Vol. 6, p. 164–178.

HOWE, G. M. 1956, The moisture balance in England and Wales; *Weather*, Vol. 11, p. 74–82.

LATHAM, J. 1966, Some electrical processes in the atmosphere; *Weather*, Vol. 21, p. 120–127.

LIGDA, M. G. H. 1951, Radar storm observation; In *Compendium of Meteorology* (*Ed.* Malone, T. F.), (American Meteorological Society, Boston, Mass.), p. 1265–1282.

LINSLEY, R. K. and FRANZINI, J. B. 1964, *Water-Resources Engineering;* (McGraw-Hill, New York), 654 pp.

LUDLAM, F. H. 1956, The structure of rainclouds; *Weather*, Vol. 11, p. 187–196.

McDONALD, J. E. 1962, The evaporation–precipitation fallacy; *Weather*, Vol. 17, p. 168–177.

MASON, B. J. 1959, Recent developments in the physics of rain and rainmaking, *Weather*, Vol. 14, p. 81–97.

MASON, B. J. 1962a, *Clouds, Rain and Rainmaking;* (Cambridge), 145 pp.

MASON, B. J. 1962b, Charge generation in thunderstorms; *Endeavour*, Vol. 21, p. 156–163.

MÖLLER, F. 1951, Vierteljahrkarten des Niederschlags für die ganze Erde; *Petermann's Geographische Mitteilungen*, 95 Jahrgang, p. 1–7.

MORE, R. J. 1967, Hydrological models and geography; In *Models in Geography* (*Ed.* Chorley, R. J. and Haggett, P.) (Methuen, London), p. 145–185.

NORDBERG, W. and PRESS, H. 1964, The 'Nimbus I' meteorological satellite; *Bull. Am. Met. Soc.*, Vol. 45, p. 684–687.

PAULHUS, J. L. H. 1965, Indian Ocean and Taiwan rainfall set new records; *Monthly Weather Review*, Vol. 93, p. 331–335.

PEARL, R. T. *et al.*, 1954, *The calculation of irrigation need;* Tech. Bull. No. 4, Min. Agric., Fish. and Food (H.M.S.O., London), 35 pp.

PENMAN, H. L. 1963, *Vegetation and Hydrology;* Tech. Comm. No. 53, Commonwealth Bureau of Soils (Harpenden), 124 pp.

REITAN, C. H. 1960, Mean monthly values of precipitable water over the United States, 1946–56; *Monthly Weather Review*, Vol. 88, p. 25–35.

SAWYER, J. S. 1956, The physical and dynamical problems of orographic rain; *Weather*, Vol. 11, p. 375–381.

SCHERMERHORN, V. P. 1967, Relations between topography and annual precipitation in western Oregon and Washington; *Water Resources Research*, Vol. 3, p. 707–711.

SIMPSON, C. G. 1941, On the formation of clouds and rain; *Quart. J. Roy. Met. Soc.*, Vol. 67, p. 99–133.

SUTCLIFFE, R. C. 1956, Water balance and the general circulation of the atmosphere; *Quart. J. Roy. Met. Soc.*, Vol. 82, p. 385–395.

WARD, R. C. 1963, Measuring potential evapotranspiration; *Geography*, Vol. 47, p. 49–55.

WEISCHET, W. 1965 Der tropische-konvective und der ausser tropische-advektive Typ der vertikalen Niederschlagsverteilung; *Erdkunde*, Vol. 19, p. 6–14.

WIDGER, W. K. 1961, Satellite meteorology—fancy and fact; *Weather*, Vol. 16, p. 47–55.

WINSTON, J. S. 1962, The operational use of meteorological satellite data; *Ann. New York Acad. Sci.*, Vol. 93, p. 775–812.

WORLD METEOROLOGICAL ORGANIZATION 1956, *International Cloud Atlas;* (Geneva).

YARNELL, D. L. 1935, Rainfall intensity–frequency data; *U.S. Dept. Agr., Misc. Pub.* 204.

Chapter 3. Atmospheric motion

BARRY, R. G. 1967, Models in meteorology and climatology; In *Models in Geography* (*Ed.* Chorley, R. J. and Haggett, P.), (Methuen, London), p. 97–144.

BERAN, W. D. 1967, Large amplitude lee waves and chinook winds; *J. Appl. Met.*, Vol. 6, p. 865–877.

BORCHERT, J. R. 1953, Regional differences in world atmospheric circulation; *Ann. Assn. Amer. Geog.*, Vol. 43, p. 14–26.

CROWE, P. R. 1950, The seasonal variation in the strength of the trades; *Trans. Inst. Brit. Geog.*, No. 16, p. 23–47.

DEFANT, F. 1951, Local winds; In *Compendium of Meteorology* (*Ed.* Malone, T. F.) (American Meteorological Society, Boston, Mass.), p. 655–672.

DEFANT, F. and TABA, H. 1957, The threefold structure of the atmosphere and the characteristics of the tropopause; *Tellus*, Vol. 9, p. 259–274.

DIETRICH, G. 1963, *General Oceanography: An Introduction;* (Wiley, New York), 588 pp.

EDDY, A. 1966, The Texas coast sea-breeze: A pilot study; *Weather*, Vol. 21, p. 162–170.

GARBELL, M. A. 1947, *Tropical and Equatorial Meteorology;* (Pitman, London), 237 pp.

GLENN, C. L. 1961, The chinook; *Weatherwise*, Vol. 14, p. 175–182.

HARE, F. K. 1965, Energy exchanges and the general circulation; *Geography*, Vol. 50, p. 229–241.

KUENEN, PH. H. 1955, *Realms of Water;* (Cleaver-Hulme Press, London), 327 pp.

LAMB, H. H. 1960, Representation of the general atmospheric circulation; *Met. Mag.*, Vol. 89, p. 319–330.

LOCKWOOD, J. G. 1962, Occurrence of föhn winds in the British Isles; *Met. Mag.*, Vol. 91, p. 57–65.

McDONALD, J. E. 1952, The Coriolis effect; *Sci. American*, Vol. 186, p. 72–78.

O'CONNOR, J. F. 1961, Mean circulation patterns based on 12 years of recent northern hemispheric data; *Monthly Weather Review*, Vol. 89, p. 211–228.

PALMÉN, E. 1951, The role of atmospheric disturbances in the general circulation; *Quart. J. Roy. Met. Soc.*, Vol. 77, p. 337–354.

PFEFFER, R. L. 1964, The global atmospheric circulation; *Trans. New York Acad. Sci.*, Ser. II, Vol. 26, p. 984–997.

RIEHL, H. 1962a, General atmospheric circulation of the tropics; *Science*, Vol. 135, p. 13–22.

RIEHL, H. 1962b, *Jet streams of the atmosphere;* Tech. Paper No. 32, Colorado State Univ., 117 pp.

RIEHL, H. *et al.* 1954, The jet stream; *Met. Monogr.*, Vol. 2, No. 7 (American Meteorological Society, Boston, Mass.), 100 pp.

ROSSBY, C-G. 1941, The scientific basis of modern meteorology; In U.S. Dept. of Agriculture Yearbook, *Climate and Man*, p. 599–655.

ROSSBY, C-G. 1949, On the nature of the general circulation of the lower atmosphere; In *The Atmosphere of the Earth and Planets* (*Ed.* Kuiper, G. P.) (Chicago), p. 16–48.

SAWYER, J. S. 1957, Jet stream features of the earth's atmosphere; *Weather*, Vol. 12, p. 333–344.

SCORER, R. S. 1958, *Natural Aerodynamics;* (Pergamon Press), 312 pp.

SCORER, R. S. 1961, Lee waves in the atmosphere; *Sci. American*, Vol. 204, p. 124–134.

STARR, V. P. 1956, The general circulation of the atmosphere; *Sci. American*, Vol. 195, p. 40–45.

TUCKER, G. B. 1961, Some developments in climatology during the last decade; *Geography*, Vol. 46, p. 198–207.

TUCKER, G. B. 1962, The general circulation of the atmosphere; *Weather*, Vol. 17, p. 320–340.

VAN ARX, W. S. 1962, *Introduction to Physical Oceanography;* (Addison-Wesley, Reading, Mass.), 422 pp.

VAN LOON, H. 1964, Mid-season average zonal winds at sea level and at 500 mb south of 25°S and a brief comparison with the northern hemisphere; *J. Appl. Met.*, Vol. 3, p. 554–563.

WACO, D. E. 1968, Frost pockets in the Santa Monica Mountains of southern California; *Weather*, Vol. 23, p. 456–461.

WALLINGTON, C. E. 1960, An introduction to lee waves in the atmosphere; *Weather*, Vol. 15, p. 269–276.

WICKHAM, P. G. 1966, Weather for gliding over Britain; *Weather*, Vol. 21, p. 154–161.

Chapter 4. Air masses, fronts, and depressions

BARRETT, E. C. 1964, Satellite meteorology and the geographer; *Geography*, Vol. 49, p. 377–386.

BATES, F. C. 1962, Tornadoes in the central United States; *Trans. Kansas Acad. Sci.*, Vol. 65, p. 215–246.

BELASCO, J. E. 1952, Characteristics of air masses over the British Isles; *Geophysical Memoirs*, Meteorological Office, Vol. 11, No. 87, 34 pp.

BOYDEN, C. J. 1960, The use of upper air charts in forecasting; *The Marine Observer*, Vol. 30, p. 27–31.

BOYDEN, C. J. 1963, Development of the jet stream and cut-off circulations; *Met. Mag.*, Vol. 92, p. 319–328.

BRUNK, I. W. 1953, Squall Lines; *Bull. Am. Met. Soc.*, Vol. 34, p. 1–9.

CRESSMAN, G. P. 1965, Numerical weather prediction in daily use; *Science*, Vol. 148, p. 319–327.

CROWE, P. R. 1949, The trade wind circulation of the world; *Trans. Inst. Brit. Geog.*, No. 15, p. 38–56.

CROWE, P. R. 1965, The geographer and the atmosphere; *Trans. Inst. Brit. Geog.*, No. 36, p. 1–19.

FAWBUSH, E. J. and MILLER, R. C. 1954, The types of airmasses in which North American tornadoes form; *Bull. Am. Met. Soc.*, Vol. 35, p. 154–165.

FREEMAN, M. H. 1961, Fronts investigated by the Meteorological Research Flight; *Met. Mag.*, Vol. 90, p. 189–203.

FULKS, J. R. 1951, The instability line; In *Compendium of Meteorology* (*Ed.* Malone, T. F.) (American Meteorological Society, Boston, Mass.), p. 647–652.

GALLOWAY, J. L. 1958a, The three-front model; its philosophy, nature, construction and use; *Weather*, Vol. 13, p. 3–10.

GALLOWAY, J. L. 1958b, The three-front model, the tropopause and the jet stream; *Weather*, Vol. 13, p. 395–403.

GALLOWAY, J. L. 1960, The three-front model, the developing depression and the occluding process; *Weather*, Vol. 15, p. 293–309.

GENTILLI, J. 1949, Air masses of the southern hemisphere; *Weather*, Vol. 4, p. 258–261 and 292–297.

GODSON, W. L. 1950, The structure of North American weather systems; *Cent. Proc. Roy. Met. Soc.* (London), p. 89–106.

HARE, F. K. 1960, The westerlies; *Geog. Rev.*, Vol. 50, p. 345–367.

HOUGHTON, D. M. 1965, Current forecasting practice; *Quart. J. Roy. Met. Soc.*, Vol. 91, p. 524–526.

KLEIN, W. H. 1948, Winter precipitation as related to the 700-mb circulation; *Bull. Am. Met. Soc.*, Vol. 29, p. 439–453.

KLEIN, W. H. 1957, Principal tracks and mean frequencies of cyclones and anticyclones in the Northern Hemisphere; *Research Paper* No. 40, *Weather Bureau* (Washington), 60 pp.

KNIGHTING, E. 1958, Numerical weather forecasting; *Weather*, Vol. 13, p. 39–50.

LAMB, H. H. 1951, Essay on frontogenesis and frontolysis; *Met. Mag.*, Vol. 80, p. 35–36, 65–71, and 97–106.

LUDLAM, F. H. 1961, The hailstorm; *Weather*, Vol. 16, p. 152–162.

MILES, M. K. 1961, The basis of present-day weather forecasting; *Weather*, Vol. 16, p. 349–363.

MILES, M. K. 1962, Wind, temperature and humidity distribution at some cold fronts over S.E. England; *Quart. J. Roy. Met. Soc.*, Vol. 88, p. 286–300.

MILLER, R. C. 1959, Tornado-producing synoptic patterns; *Bull. Am. Met. Soc.*, Vol. 40, p. 465–472.

MILLER, R. C. and STARRETT, L. G. 1962, Thunderstorms in Great Britain; *Met. Mag.* Vol. 91, p. 247–255.

PENNER, C. M. 1955, A three-front model for synoptic analyses; *Quart. J. Roy. Met. Soc.*, Vol. 81, p. 89–91.

PETTERSSEN, S. 1950, Some aspects of the general circulation of the atmosphere; *Cent. Proc. Roy. Met. Soc.* (London), p. 120–155.

POTHECARY, I. J. W. 1956, Recent research on fronts; *Weather*, Vol. 12, p. 147–150.

REED, R. J. 1960, Principal frontal zones of the northern hemisphere in winter and summer; *Bull. Am. Met. Soc.*, Vol. 41, p. 591–598.

RICHTER, D. A. and DAHL, R. A. 1958, Relationship of heavy precipitation to the jet maximum in the eastern United States, *Monthly Weather Review*, Vol. 86, p. 368–376.

SAWYER, J. S. 1967, Weather forecasting and its future; *Weather*, Vol. 22, p. 350–359 and 400–408.

SHOWALTER, A. K. 1939, Further studies of American air mass properties; *Monthly Weather Review*, Vol. 67, p. 204–218.

SUTCLIFFE, R. C. 1964, Weather forecasting by electronic computer; *Endeavour*, Vol. 23, p. 27–32.

SUTCLIFFE, R. C. and FORSDYKE, A. G. 1950, The theory and use of upper air thickness patterns in forecasting; *Quart. J. Roy. Met. Soc.*, Vol. 76, p. 189–217.

VEDERMAN, J. 1954, The life cycles of jet streams and extratropical cyclones; *Bull. Am. Met. Soc.*, Vol. 35, p. 239–244.

WALLINGTON, C. E. 1963, Mesoscale patterns of frontal rainfall and cloud; *Weather*, Vol. 18, p. 171–181.

YOSHINO, M. M. 1967, Maps of the occurrence frequencies of fronts in the rainy season in early summer over east Asia; *Science Reports of the Tokyo University of Education*, Section C, Vol. 9, No. 89, p. 211–245.

Chapter 5. Weather and climate in temperate latitudes

BAILEY, H. P. 1964, Toward a unified concept of the temperate climate; *Geog. Rev.*, Vol. 54(4), p. 516–545.

BARRY, R. G. 1963, Aspects of the synoptic climatology of central south England; *Met. Mag.*, Vol. 92, p. 300–308.

BARRY, R. G. 1967a, Seasonal location of the arctic front over North America; *Geog. Bull.*, Vol. 9, p. 79–95.

BARRY, R. G. 1967b, The prospect for synoptic climatology: A case study; In Steel, R. W. and Lawton, R. (*Eds.*), *Liverpool Essays in Geography* (Longmans, London), p. 85–106.

BELASCO, J. E. 1948, The incidence of anticyclonic days and spells over the British Isles; *Weather*, Vol. 3, p. 233–242.

BILHAM, E. G. 1938, *The Climate of the British Isles;* (Macmillan, London), 347 pp.

BLEEKER, W. and ANDRE, M. J. 1951, On the diurnal variation of precipitation, particularly over central U.S.A., and its relation to large-scale orographic circulation systems; *Quart. J. Roy. Met. Soc.*, Vol. 77, p. 260–277.

BORCHERT, J. 1950, The climate of the central North American grassland; *Ann. Assn. Am. Geog.*, Vol. 40, p. 1–39.

BRYSON, R. A. 1966, Air masses, stream lines and the boreal forest; *Geog. Bull.* Vol. 8, p. 228–269.

BRYSON, R. A. and LAHEY, J. F. 1958, *The March of the Seasons;* (Meteorological Department, University of Wisconsin), 41 pp.

BRYSON, R. A. and LOWRY, W. P. 1955, Synoptic climatology of the Arizona summer precipitation singularity; *Bull. Am. Met. Soc.*, Vol. 36, p. 329–339.

BURBIDGE, F. E. 1951, The modification of continental polar air over Hudson Bay; *Quart. J. Roy. Met. Soc.*, Vol. 77, p. 365–374.

BUTZER, K. W. 1960, Dynamic climatology of large-scale circulation patterns in the Mediterranean area; *Meteorologische Rundschau*, Vol. 13, p. 97–105.

ENVIRONMENTAL SCIENCES SERVICES ADMINISTRATION 1968, *Climatic Atlas of the United States;* U.S. Dept. of Commerce (Washington, D.C.), 80 pp.

GORCZYNSKI, W. 1920, Sur le calcul du degré du contintentalisme et son application dans la climatologie; *Geografiska Annaler*, Vol. 2, p. 324–331.

GREEN, C. R. and SELLERS, W. D. 1964, *Arizona Climate;* (University of Arizona Press, Tucson), 503 pp.

HARE, F. K. 1950, Some climatological problems of the Arctic and Subarctic; In *Compendium of Meteorology* (*Ed.* Malone, T. F.) (American Meteorological Society, Boston, Mass.), p. 952–963.

HAWKE, E. L. 1933, Extreme diurnal ranges of air temperature in the British Isles; *Quart. J. Roy. Met. Soc.*, Vol. 59, p. 261–265.

HORN, L. H. and BRYSON, R. A. 1960, Harmonic analysis of the annual march of precipitation over the United States; *Ann. Assoc. Am. Geog.*, Vol. 50, p. 157–171.

HUTTARY, J. 1950, Die Verteilung der Niederschläge auf die Jahreszeiten im Mittelmeergebiet; *Meteorologische Rundschau*, Vol. III, p. 111–119.

KENDREW, W. G. and CURRIE, B. W. 1955, *The Climate of Central Canada;* (Ottawa), 194 pp.

KLEIN, W. H. 1963, Specification of precipitation from the 700-mb circulation; *Monthly Weather Review*, Vol. 91, p. 527–536.

LAMB, H. H. 1950, Types and spells of weather around the year in the British Isles: Annual trends, seasonal structure of the year, singularities; *Quart. J. Roy. Met. Soc.*, Vol. 76, p. 393–438.

LONGLEY, R. W. 1967, The frequency of Chinooks in Alberta; *The Albertan Geographer*, No. 3, p. 20–22.

LUMB, F. E. 1961, Seasonal variations of the sea surface temperature in coastal waters of the British Isles; *Met. Office Sci. Paper No. 6*, M.O. 685, 21 pp.

MANLEY, G. 1944, Topographical features and the climate of Britain; *Geog. J.*, Vol. 103, p. 241–258.

MANLEY, G. 1945, The effective rate of altitude change in temperate Atlantic climates; *Geog. Rev.*, Vol. 35, p. 408–417.

METEOROLOGICAL OFFICE 1952, *Climatological Atlas of the British Isles;* M.O. 488 (H.M.S.O., London), 139 pp.

METEOROLOGICAL OFFICE 1962, *Weather in the Mediterranean;* (H.M.S.O., London), Vol. I, General Meteorology, 2d ed., 362 pp.

METEOROLOGICAL OFFICE 1964a, *Weather in the Mediterranean;* M.O. 391b (H.M.S.O., London), Vol. 2, 372 pp.

METEOROLOGICAL OFFICE 1964b, *Weather in Home Fleet Waters;* M.O. 732a (H.M.S.O., London), Vol. 1, The Northern Seas, Part 1, 265 pp.

NAMIAS, J. 1964, Seasonal persistence and recurrence of European blocking during 1958–1960; *Tellus,* Vol. 16, p. 394–407.

RAYNER, J. N. 1961, *Atlas of Surface Temperature Frequencies for North America and Greenland;* (Arctic Meteorological Research Group, McGill University, Montreal).

REX, D. F. 1950–51, The effect of Atlantic blocking action upon European climate; *Tellus,* Vol. 2, p. 196–211 and 275–301, Vol. 3, p. 100–111.

SHAW, E. M. 1962, An analysis of the origins of precipitation in Northern England, 1956–60; *Quart. J. Roy. Met. Soc.,* Vol. 88, p. 539–547.

SIVALL, T. 1957, Sirocco in the Levant; *Geografiska Annaler,* Vol. 39, p. 114–142.

SUMNER, E. J. 1959, Blocking anticyclones in the Atlantic-European sector of the northern hemisphere; *Met. Mag.,* Vol. 88, p. 300–311.

THOMAS, M. K. 1964, A survey of Great Lakes snowfall; *Great Lakes Research Division, Univ. of Michigan, Publication No.* 11, p. 294–310.

THORNTHWAITE, C. W. and MATHER, J. R. 1955, *The Moisture Balance;* Publications in Climatology, Vol. 8, No. 1 (Laboratory of Climatology, Centerton, N.J.), 104 pp.

UNITED STATES WEATHER BUREAU 1947, *Thunderstorm Rainfall;* (Vicksburg, Mississippi), 331 pp.

VILLMOW, J. R. 1956, The nature and origin of the Canadian dry belt; *Ann. Assn. Am. Geog.,* Vol. 46, p. 211–232.

VISHER, S. S. 1954, *Climatic Atlas of the United States;* (Harrd), 403 pp.

WALLÉN, C. C. 1960, Climate; In *The Geography of Norden* (Ed. Sømme, A.) (Cappelens Forlag, Oslo), p. 41–53.

Chapter 6. Tropical weather and climate

ACADEMICA SINICA 1957–58, On the general circulation over eastern Asia; *Tellus,* Vol. 9, p. 432–446, Vol. 10, p. 58–75 and 299–312.

BECKINSALE, R. P. 1957, The nature of tropical rainfall; *Tropical Agriculture;* Vol. 34, p. 76–98.

BERGERON, T. 1954, The problem of tropical hurricanes; *Quart. J. Roy. Met. Soc.,* Vol. 80, p. 131–164.

BLUMENSTOCK, D. I. 1958, Distribution and characteristics of tropical climates; *Proc. 9th Pacific Sci. Congr.,* Vol. 20, p. 3–23.

CHANG, J.-H. 1957, Air mass maps of China proper and Manchuria; *Geography,* Vol. 42, p. 142–148.

CHANG, J.-H. 1962, Comparative climatology of the tropical western margins of the northern oceans; *Ann. Assn. Am. Geog.,* Vol. 52, p. 221–227.

CHANG, J.-H. 1967, The Indian summer monsoon; *Geog. Rev.,* Vol. 57, p. 373–396.

CROWE, P. R. 1949, The Trade Wind circulation of the world; *Trans. Inst. Brit. Geog.,* No. 15, p. 37–56.

CROWE, P. R. 1951, Wind and weather in the equatorial zone; *Trans. Inst. Brit. Geog.*, No. 17, p. 23–76.

CRY, G. W. 1965, Tropical cyclones of the North Atlantic Ocean; *Tech. Paper* No. 55, *Weather Bureau* (Washington), 148 pp.

CURRY, L. and ARMSTRONG, R. W. 1959, Atmospheric circulation of the tropical Pacific ocean; *Geografiska Annaler*, Vol. 41, p. 245–255.

DUNN, G. E. and MILLER, B. I. 1960, *Atlantic Hurricanes;* (Louisiana State University Press), 326 pp.

ELDRIDGE, R. H. 1957, A synoptic study of West African disturbance lines; *Quart. J. Roy. Met. Soc.*, Vol. 83, p. 303–314.

FLOHN, H. 1968, *Contributions to a meteorology of the Tibetan Highlands;* Atmos. Sci. Paper No. 130 (Colorado State Univ., Fort Collins), 120 pp.

FOSBERG, F. R., GARNIER, B. J., and KÜCHLER, A. W. 1961, Delimitation of the humid tropics; *Geog. Rev.*, Vol. 51, p. 333–347.

FROST, R. and STEPHENSON, P. M. 1965, Mean streamlines and isotachs at standard pressure levels over the Indian and west Pacific Oceans and adjacent land areas; *Geophys. Mem.*, Vol. 14 (No. 109), (H.M.S.O., London), 24 pp.

GARBELL, M. A. 1947, *Tropical and Equatorial Meteorology;* (Pitman, London), 237 pp.

GARNIER, B. J. 1967, Weather conditions in Nigeria; *Climatological Research Series No 2*, McGill Univ., Montreal, 163 pp.

GRAY, W. M. 1968, Global view of the origin of tropical disturbances and hurricanes; *Monthly Weather Review*, Vol. 96, p. 669–700.

GREGORY, S. 1965, *Rainfall over Sierra Leone;* Geography Department, University of Liverpool, Research Paper No. 2, 58 pp.

HUTCHINGS, J. W. (*Ed.*) 1964, *Proceedings of the Symposium on Tropical Meteorology;* (New Zealand Meteorological Service, Wellington), 737 pp.

INDIAN METEOROLOGICAL DEPARTMENT 1960, *Monsoons of the World;* (Delhi), 270 pp.

JORDAN, C. L. 1955, Some features of the rainfall at Guam; *Bull. Am. Met. Soc.*, Vol. 36, p. 446–455.

KOTESWARAM, P. 1958, The easterly jet stream in the tropics; *Tellus*, Vol. 10, p. 43–57.

LOCKWOOD, J. G. 1965, The Indian monsoon—a review; *Weather*, Vol. 20, p. 2–8.

LOGAN, R. F., 1960, The Central Namib Desert, South West Africa; *National Academy of Sciences, National Research Council, Publication* 758 (Washington, D.C.), 162 pp.

LOWELL, W. E., 1954, Local weather of the Chicama Valley, Peru; *Archiv für Meteorologie, Geophysik und Bioklimatologie*, Series B, Vol. 5, p. 41–51.

LYDOLPH, P. E. 1957, A comparative analysis of the dry western littorals; *Ann. Assn. Am. Geog.*, Vol. 47, p. 213–230.

MAEJIMA, I. 1967, Natural seasons and weather singularities in Japan; *Geog. Rept. No. 2, The Tokyo Metropolitan University*, p. 77–103.

MALKUS, J. S. 1955–56, The effects of a large island upon the trade-wind air stream; *Quart. J. Roy. Met. Soc.*, Vol. 81, p. 538–550, and Vol. 82, p. 235–238.

MALKUS, J. S. 1958, Tropical weather disturbances: why do so few become hurricanes?; *Weather*, Vol. 13, p. 75–89.

MALKUS, J. S. and RIEHL, H. 1964, *Cloud Structure and Distributions over the Tropical Pacific Ocean;* (University of California Press, Berkeley and Los Angeles), 229 pp.

MILLER, B. I. 1967, Characteristics of hurricanes; *Science*, Vol. 157, p. 1389–1399.

MINK, J. F. 1960, Distribution pattern of rainfall in the leeward Koolau Mountains, Oahu, Hawaii; *Jour. Geophys. Res.*, Vol. 65, p. 2869–2876.

PALMÉN, E. 1948, On the formation and structure of tropical hurricanes; *Geophysica*, No. 3, p. 26–38.

PALMER, C. E. 1951, Tropical meteorology; In *Compendium of Meteorology* (*Ed.* Malone, T. F.) (American Meteorological Society, Boston, Mass.), p. 859–880.

PÉDELABORDE, P. 1958, *Les Moussons;* (Armand Colin, Paris), 208 pp. (English ed., Methuen, London, 1963, 196 pp.).

RAGHARAN, K. 1967, Influence of tropical storms on monsoon rainfall in India; *Weather*, Vol. 22, p. 250–255.

RAMAGE, C. S. 1952, Relationships of general circulation to normal weather over southern Asia and the western Pacific during the cool season; *J. Met.*, Vol. 9, p. 403–408.

RAMAGE, C. S. 1964, Diurnal variation of summer rainfall in Malaya; *J. Trop. Geog.*, Vol. 19, p. 62–68.

RAMAGE, C. S. 1968, Problems of a monsoon ocean; *Weather*, Vol. 23, p. 28–36.

RAMASWAMY, C. 1956, On the sub-tropical jet stream and its role in the development of large-scale convection; *Tellus*, Vol. 8, p. 26–60.

RAMASWAMY, C. 1962, Breaks in the Indian summer monsoon as a phenomenon of interaction between the easterly and the sub-tropical westerly jet streams; *Tellus*, Vol. 14, p. 337–349.

REITER, E. R. and HEUBERGER, H. 1960, A synoptic example of the retreat of the Indian summer monsoon; *Geografiska Annaler*, Vol. 42, p. 17–35.

RIEHL, H. 1954, *Tropical Meteorology;* (McGraw-Hill, New York), 392 pp.

RIEHL, H. 1962, General atmospheric circulation in the tropics; *Science*, Vol. 135, p. 13–22.

RIEHL, H. 1963, On the origin and possible modification of hurricanes; *Science*, Vol. 141, p. 1001–1010.

SAITO, R. 1959, The climate of Japan and her meteorological disasters; *Proceedings of the International Geophysical Union, Regional Conference in Japan* (*Tokyo*), p. 173–183.

SAWYER, J. S. 1947, The structure of the Intertropical Front over N.W. India during the S.W. Monsoon; *Quart. J. Roy. Met. Soc.*, Vol. 73, p. 346–369.

SCORER, R. S. 1966, Origin of cyclones; *Science Journal*, Vol. 2(3), p. 46–52.

THOMPSON, B. W. 1951, An essay on the general circulation over South-East Asia and the West Pacific; *Quart. J. Roy. Met. Soc.*, Vol. 77, p. 569–597.

THOMPSON, B. W. 1965, *The Climate of Africa* (Atlas); (O.U.P., Nairobi).

TREWARTHA, G. T. 1958, Climate as related to the jet stream in the Orient; *Erdkunde*, Vol. 12, p. 205–214.

WATTS, I. E. M. 1955, *Equatorial Weather, with particular reference to Southeast Asia;* (University of London Press), 186 pp.

YIN, M. T. 1949, A synoptic-aerologic study of the onset of the summer monsoon over India and Burma; *J. Met.*, Vol. 6, p. 393–400.

Chapter 7. Urban and forest climates

ATKINSON, B. W. 1968, A preliminary examination of the possible effect of London's urban area on the distribution of thunder rainfall 1951–60; *Trans. Inst. Brit. Geog.*, No. 44, p. 97–118.

CABORN, J. M. 1955, The influence of shelter-belts on microclimate; *Quart. J. Roy. Met. Soc.*, Vol. 81, p. 112–115.

CHAGNON, S. A., JR. 1969, Recent studies of urban effects on precipitation in the United States; *Bull. Am. Met. Soc.*, Vol. 50, p. 411–421.

CHANDLER, T. J. 1965, *The Climate of London;* (Hutchinson, London), 292 pp.

CHANDLER, T. J. 1967, Absolute and relative humidities in towns; *Bull. Am. Met. Soc.*, Vol. 48, p. 394–399.

COMMITTEE ON AIR POLLUTION 1955, *Report;* (Cmd 9322, H.M.S.O., London).

COUTTS, J. R. H. 1955, Soil temperatures in an afforested area in Aberdeenshire; *Quart. J. Roy. Met. Soc.*, Vol. 81, p. 72–79.

DUCKWORTH, F. S. and SANDBERG, J. S. 1954, The effect of cities upon horizontal and vertical temperature gradients; *Bull. Am. Met. Soc.*, Vol. 35, p. 198–207.

FOOD AND AGRICULTURE ORGANISATION OF THE UNITED NATIONS 1962, *Forest Influences;* Forestry and Forest Products Studies No. 15 (Rome), 307 pp.

GARNETT, A. 1957, Climate, relief and atmospheric pollution in the Sheffield region; *The Advancement of Science*, Vol. 13, p. 331–341.

HEWSON, E. W. 1951, Atmospheric pollution; In *Compendium of Meteorology (Ed.* Malone, T. F.) (American Meteorological Society, Boston, Mass.), p. 1139–1157.

KITTREDGE, J. 1948, *Forest Influences;* (McGraw-Hill, New York), 394 pp.

LANDSBERG, H. E. 1956, The climate of towns; In *Man's Role in Changing the Face of the Earth (Ed.* Thomas, W. L.) (Chicago), p. 584–603.

MacDONALD, G. J. F. *(Chairman)* 1966, Weather and climate modification: Problems and prospects; *Bull. Am. Met. Soc.*, Vol. 47, p. 4–19.

MacDONALD, G. J. F. 1968, Weather modification; *Science Journal*, Vol. 4, p. 39–44.

MARSHALL, W. A. L. 1952, *A Century of London Weather;* Met. Office, Air Ministry, Rept M.O. 508 (H.M.S.O., London), 103 pp.

MEETHAM, A. R. 1952, *Atmospheric Pollution;* (Pergamon, London), 268 pp.

MEETHAM, A. R. 1955, Know your fog; *Weather*, Vol. 10, p. 103–105.

NATIONAL ACADEMY OF SCIENCES 1966, *Spacecraft in Geographic Research;* National Research Council, Publication 1353 (Washington), p. 23–38.

OLGYAY, V. 1963, *Design with Climate; Bioclimatic approach to architectural regionalism;* (Princeton), 190 pp.

PARRY, M. 1966, The urban 'heat island'; In *Biometeorology*, Vol. 2 *(Ed.* Tromp, S. W. and Weihe, W. H.) (Pergamon Press, London), p. 616–624.

RAMDAS, L. A. 1957, Natural and artificial modification of microclimate; *Weather*, Vol. 12, p. 237–240.

REYNOLDS, E. R. C. and LEYTON, L. 1963, Measurement and significance of through-fall in forest stands; In *The Water Relations of Plants (Ed.* Whitehead, F. M. and Rutter, A. J.) (Blackwell Scientific Publications, Oxford), p. 127–141.

RICHARDS, P. W. 1952, *The Tropical Rain Forest;* (Cambridge), 450 pp.

RUTTER, A. J. 1967, Evaporation in forests; *Endeavour*, Vol. 26, No. 97, p. 39–43.

SARGENT, F. 1967, A dangerous game: Taming the weather; *Bull. Am. Met. Soc.*, Vol. 48, p. 452–458.

SCORER, R. 1968, *Air Pollution;* (Pergamon, Oxford and London), 151 pp.

SEWELL, W. R. D. *(Ed.)* 1966, Human dimensions of weather modification; *Univ. Chicago, Dept. of Geography, Research Paper* 105, 423 pp.

SIMPSON, R. H. and SIMPSON, J. 1966, 'Why experiment on tropical hurricanes?' *Trans. New York Acad. Sci.*, Ser. II, Vol. 28, No. 8, p. 1045–1062.

SOPPER, W. E. and LULL, H. W. *(Eds.)*, 1967, *International Symposium on Forest Hydrology;* (Pergamon, Oxford and London), 813 pp.

SUKACHEV, V. and DYLIS, N. 1968, *Fundamentals of Forest Biogeocoenology;* (Oliver and Boyd, Edinburgh), 672 pp.

TURNER, W. C. 1955, Atmospheric pollution; *Weather*, Vol. 10, p. 110–119.

ZON, R. 1941, Climate and the nation's forests; U.S. Dept. of Agriculture Yearbook *Climate and Man*, p. 477–498.

Chapter 8. Climatic variability, trends, and fluctuations

AHLMANN, H. W. 1948, The present climatic fluctuation; *Geog. J.*, Vol. 113, p. 165–193.

BECKINSALE, R. P. 1965, Climatic change: A critique of modern theories; In *Essays in Geography for Austin Miller* (*Ed.* Whittow, J. B. and Wood, P. D.) (University of Reading Press), p. 1–38.

CALLENDAR, G. S. 1961, Temperature fluctuations and trends over the earth; *Quart. J. Roy. Met. Soc.*, Vol. 87, p. 1–12.

CONRAD, V. and POLLAK, L. W. 1950, *Methods in Climatology;* (Harvard). See Chapter 2, Statistical analysis of climatic elements, p. 17–60.

DORF, E. 1960, Climatic changes of the past and present; *Am. Scientist*, Vol. 48, p. 341–364.

GREGORY, S. 1962, *Statistical Methods and the Geographer;* (Longmans, London), 240 pp.

KIMBLE, J. H. T. 1950, The changing climate; *Sci. American*, Vol. 182, p. 48–53.

KRAUS, E. B. 1955, Secular changes of tropical rainfall regimes; *Quart. J. Roy. Met. Soc.*, Vol. 81, p. 198–210.

LAMB, H. H. 1959, The changing climate; *The New Scientist*, Vol. 6, p. 740–744.

LAMB, H. H. 1961, Fundamentals of climate; In *Descriptive Palaeoclimatology* (*Ed.* Nairn, A. E. M.) (Interscience Publishers, New York), p. 8–44.

LAMB, H. H. 1962, Changes of climate before and since the Industrial Revolution; *Research*, Vol. 15, p. 501–509.

LAMB, H. H. 1963, What can we find out about the trend of our climate?; *Weather*, Vol. 18, p. 194–206.

LAMB, H. H. 1965, Frequency of weather types; *Weather*, Vol. 20, p. 9–12.

LAMB, H. H. and JOHNSON, A. I. 1959, Climatic variation and observed changes in the general circulation; *Geografiska Annaler*, Vol. 41, p. 94–134.

LAWRENCE, E. N. 1965, Terrestrial climate and the solar cycle; *Weather*, Vol. 20, p. 334–343.

LEOPOLD, L. B. 1951, Rainfall frequency: An aspect of climatic variation; *Trans. Am. Geophys. Union*, Vol. 32(3), p. 347–357.

LEWIS, P. 1960, The use of moving averages in the analysis of time-series; *Weather*, Vol. 15, p. 121–126.

LYSGAARD, L. 1949, Recent climatic fluctuations; *Folia Geographica Danica, Kongelige Dansk Geog.*, Selskab, 5, p. 215.

MANLEY, G. 1958, Temperature trends in England, 1698–1957; *Archiv für Meteorologie, Geophysik und Bioklimatologie*, Series B (Vienna), Vol. 9, p. 413–433.

MATHER, J. R. 1954, The present climatic fluctuation and its bearing on a reconstruction of Pleistocene climatic conditions; *Tellus*, Vol. 3, p. 287.

MITCHELL, J. M., JR. (*Ed.*) 1968, Causes of climatic change; *Met. Monogr.*, Vol. 8, No. 30 (American Meteorological Society, Boston, Mass.), 160 pp.

PLASS, G. N. 1959, Carbon dioxide and climate; *Sci. American*, Vol. 201, p. 41–47.

SHAPLEY, H. (*Ed.*) 1953, *Climatic Change;* (Harvard), 318 pp.

TUCKER, G. B. 1964, Solar influences on the weather; *Weather*, Vol. 19, p. 302–311.

UNESCO, 1963, Changes of Climate; *Arid Zone Research*, Vol. 20 (UNESCO, Paris), 488 pp.

WEXLER, H. 1956, Variations in insolation, general circulation and climate; *Tellus*, Vol. 8, p. 480–494.

WILLETT, H. C. 1950, Temperature trends of the past century; *Cent. Proc. Roy. Met. Soc.* (London), p. 195–206.

APPENDIX 1
Climatic Classification

The purpose of any classification system is to obtain an efficient arrangement of information in a simplified and generalized form. Thus, climatic statistics can be organized in order to describe and delimit the major types of climate in quantitative terms. Obviously no single classification can serve more than a limited number of purposes satisfactorily and many different schemes have therefore been developed. Some schemes merely provide a convenient nomenclature system, whereas others are an essential preliminary to further study. Many climatic classifications, for instance, are concerned with the relationships between climate and vegetation or soils, but surprisingly few attempts have been made to base a classification on the direct effects of climate on man.

Only the basic principles of the three groups of the most widely known classification systems are summarized here. Further information may be found in the listed references.

A. GENERIC CLASSIFICATIONS RELATED TO PLANT GROWTH OR VEGETATION

The numerous schemes which have been suggested for relating climatic limits to plant growth or vegetation groups rely on two basic criteria—the degree of aridity and of warmth.

Aridity is not simply a matter of low precipitation, but of the effective precipitation (that is, precipitation minus evaporation). The ratio of rainfall/temperature has been used as such an index of precipitation effectiveness, on the grounds that higher temperatures increase evaporation. The ratio r/t was proposed by R. Lang in 1915 (where r = mean annual rainfall in mm, and t = mean annual temperature in °C), such that $r/t < 40$ is considered arid and $r/t > 160$ perhumid.

The work of W. Köppen is the prime example of this type of classification. Between 1900 and 1936 he published several classification schemes involving considerable complexity in their full detail. Nevertheless, the system has been used extensively in geographical teaching. The key features of Köppen's final classification are:

Temperature criteria: Five of the six major climatic types are recognized on the basis of monthly mean temperature.

A—Tropical rainy climate. Coldest month $> 18°C$ (64.4°F).
B—Dry climates.
C—Warm temperate rainy climates. Coldest month between $-3°$ and 18°C, warmest month $> 10°C$ (50°F).
D—Cold boreal forest climates. Coldest month $< -3°C$ (26.6°F),* warmest month $> 10°C$.
E—Tundra climate. Warmest month 0°–10°C.
F—Perpetual frost climate. Warmest month $< 0°C$.

The arbitrary temperature limits stem from a variety of criteria, the supposed significance of the selected values being as follows: the 10°C summer isotherm correlates with the poleward limit of tree growth; the 18°C winter isotherm is critical for certain tropical plants; and the $-3°C$ isotherm indicates a few weeks of snow cover. However, these correlations are far from precise! The criteria were determined from a study of vegetation groups defined on a physiological basis (that is, according to the internal functions of plant organs) by De Candolle in 1874.

Aridity criteria:

	Steppe (BS)/Desert (BW) boundary	Forest/Steppe boundary
Winter precipitation maximum	$r/t = 1$	$r/t = 2$
Precipitation evenly distributed	$r/(t+7) = 1$	$r/(t+7) = 2$
Summer precipitation maximum	$r/(t+14) = 1$	$r/(t+14) = 2$

where
$$r = \text{annual precipitation (in cm)}$$
$$t = \text{mean annual temperature (in °C)}.$$

The criteria imply that, with winter precipitation, arid (desert) conditions occur where $r/t < 1$, semi-arid conditions where $1 < r/t < 2$. If the rain

---

* Note that many American workers use a modified version with 0°C as the C/D boundary.

falls in summer a larger amount is required to offset evaporation and maintain an equivalent total of effective precipitation.

Subdivisions of each major category are made with reference, firstly, to the seasonal distribution of precipitation (the most common of which are: f = no dry season; m = monsoonal, with a short dry season and heavy rains during the rest of the year; s = summer dry season; w = winter dry season) and, secondly, to additional temperature characteristics. Fig. Appx. 1.1A illustrates the distribution of the major Köppen climatic types on a hypothetical continent of low and uniform elevation.

A somewhat similar scheme has been proposed more recently by A. A. Miller (1951), using the following criteria:

$$\text{Boundary of arid conditions:} \quad r/t = 1/5.$$
$$\text{Boundary of semi-arid conditions:} \; r/t = 1/3,$$

where

$$r = \text{mean annual rainfall (in in.)}$$
$$t = \text{mean annual temperature (in °F)}.$$

The thermal limits relate to the *accumulated temperature*, which Miller estimated by using month-degrees—the excess of mean monthly temperatures above 43°F(6°C)—rather than the more usual day-degrees based on daily means.

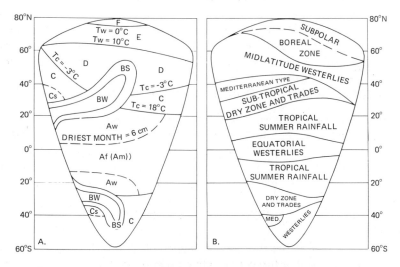

Fig. Appx. 1.1. A. The distribution of the major Köppen climatic types on a hypothetical continent of low and uniform elevation. B. The distribution of Flohn's climatic types on a hypothetical continent of low and uniform elevation (from Flohn, 1950).

C. W. Thornthwaite introduced a complex, empirical classification in 1931. An expression for *precipitation efficiency* was obtained by relating measurements of pan evaporation to temperature and precipitation. For each month the ratio

$$11.5(r/t - 10)^{10/9},$$

where

r = mean monthly rainfall (in in.)
t = mean monthly temperature (in °F)

is calculated. The sum of the twelve monthly ratios gives the precipitation efficiency (P-E) index. By determining boundary values for the major vegetation regions the following humidity provinces were defined:

		P-E index
A	Rain forest	>127
B	Forest	64–127
C	Grasslands	32–63
D	Steppe	16–31
E	Desert	<16

The second element of the classification is an index of thermal efficiency (T-E), expressed by the positive departure of monthly mean temperatures from freezing point. The index is thus the annual sum of $(t - 32/4)$ for each month. On this scale zero is frost climate and over 127 is tropical. Unlike Köppen, Thornthwaite makes moisture the primary classificatory factor for a T-E index of over 31 (the taiga/cool temperate boundary). Maps of the distribution of these climatic provinces in North America and over the world have been published, but the classification is now largely of historical interest.

B. RATIONAL, MOISTURE BUDGET CLASSIFICATIONS

Thornthwaite's most important contribution is his second (1948) classification. It is based on the concept of potential evapotranspiration and the moisture budget (see Chapter 2, Section A and Chapter 5, Section B.3.*c*). The potential evapotranspiration (PE) is calculated from the mean monthly temperature (in °C), with corrections for day length. For a 30-day month (12-hour days):

$$PE \text{ (in cm)} = 1.6(10t/I)^a$$

where

I = the sum for 12 months of $(t/5)^{1.514}$
a = a further complex function of I.

Tables have been prepared for the easy computation of these factors.

The monthly water surplus (S) or deficit (D) is determined from a moisture budget assessment, taking into account stored soil moisture. A moisture index (Im) is given by:

$$Im = (100S - 60D)/PE$$

The weighting of a deficit by 0.6 was supposed to allow for the beneficial action of a surplus in one season when moisture is stored in the subsoil, to be drawn on during subsequent droughts by deep-rooted perennials. In 1955 this weighting factor was omitted since it was recognized that a deficit can begin as soon as any moisture is removed from the soil by evaporation. The later revision also allows for a variable soil moisture storage according to vegetation cover and soil type, and permits the evaporation rate to vary with the actual soil moisture content.

A novel feature of the system is that the thermal efficiency is derived from the PE value because this itself is a function of temperature. The climatic types defined by these two factors are:

Im (1955 system)*		PE		
		cm	in.	
>100	Perhumid (A)	>114	>44.9	Megathermal (A′)
20 to 100	Humid (B₁ to B₄)	57 to 114	22.4 to 44.9	Mesothermal (B′₁ to B′₄)
0 to 20	Moist Subhumid (C₂)	28.5 to 57	11.2 to 22.4	Microthermal (C′₁ to C′₂)
−33 to 0	Dry Subhumid (C₁)	14.2 to 28.5	5.6 to 11.2	Tundra (D′)
−67 to −33	Semi-arid (D)	< 14.2	< 5.6	Frost (E′)
−100 to −67	Arid (E)			

* $Im = 100(S - D)/PE$ is equivalent to $100(r/PE - 1)$, where r = annual precipitation.

Both elements are subdivided according to the season of moisture deficit or surplus and the seasonal concentration of thermal efficiency.

The system has been applied to many regions, although no world map has yet been published. In tropical and semi-arid areas the method is not very satisfactory, but in eastern North America, for example, vegetation boundaries have been shown to coincide reasonably closely with particular PE values. This classification, unlike that of 1931, Köppen's, and many others, does not use vegetation boundaries to determine climatic ones.

M. I. Budyko in the Soviet Union has developed a similar, but more fundamental, approach using net radiation rather than temperature (see Chapter 2, Section A). He relates the net radiation available for evaporation from a wet surface (Rn) to the heat required to evaporate the mean annual precipitation (Lr). This ratio Rn/Lr (where L = latent heat of evaporation) is called the *radiational index of dryness*. It has a value less than unity in humid areas and greater than unity in dry areas. Boundary values are:

Rn/Lr	
>3.0	Desert
2.0–3.0	Semi-desert
1.0–2.0	Steppe
0.33–1.0	Forest
<0.33	Tundra

By way of comparison with the revised Thornthwaite index (Im = 100 (*r*/PE − 1)) it may be noted that Im = 100(Lr/Rn − 1) if all the net radiation is used for evaporation from the wet surface (that is, none is transferred into the ground by conduction or into the air as sensible heat). A general world map of Rn/Lr has appeared but over large parts of the earth there are as yet no measurements of net radiation.

C. GENETIC CLASSIFICATIONS

The genetic basis of large-scale (or macro-) climates is the atmospheric circulation, and this can be related to regional climatology in terms of wind regimes or air masses. One attempt, made by A. Hettner in 1931, incorporated the wind system, continentality, rainfall amount and duration, position relative to the sea, and elevation. A very generalized scheme using air masses, according to their seasonal dominance, was put forward by B. P. Alissov in 1936.

A more satisfactory system, however, was proposed in 1950 by H. Flohn. His major categories, which are based on the global wind belts and the precipitation characteristics, are as follows:

(1) Equatorial westerly zone	Constantly wet
(2) Tropical zone, winter trades	Summer rainfall
(3) Subtropical dry zone (trades or subtropical high pressure)	Dry conditions prevail
(4) Subtropical winter-rain zone (Mediterranean type)	Winter rainfall
(5) Extratropical westerly zone	Precipitation throughout the year
(6) Subpolar zone	Limited precipitation throughout the year
(6a) Boreal, continental subtype	Summer rainfall; limited winter snowfall
(7) High polar zone	Meager precipitation; summer rainfall, early winter snowfall

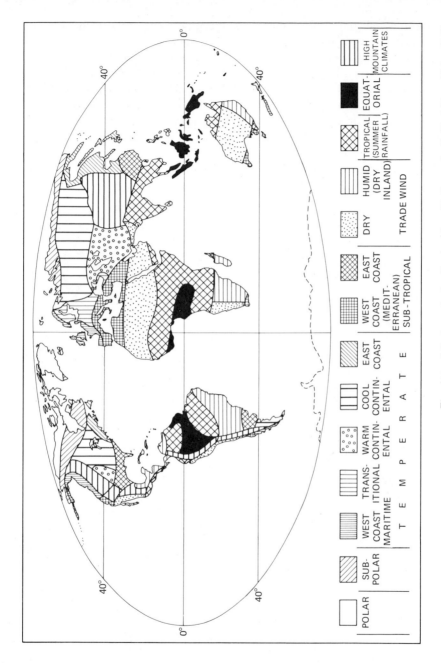

Fig. Appx. 1.2. A genetic classification of world climates by E. Neef (from Flohn, 1957).

It will be noted that temperature does not appear explicitly in the scheme. Fig. Appx. 1.1B shows the distribution of these types on a hypothetical continent. Rough general agreement between these types and those of Köppen's scheme is apparent. Note that the boreal subtype is restricted to the northern hemisphere and that the subtropical zones do not occur on the east side of a land mass. Flohn's approach has much to commend it as an introductory teaching outline. Although no world map of the distribution of these zones has been published, two maps prepared along similar lines by E. Neef and E. Kupfer were presented and discussed by Flohn in 1957. Neef's map is reproduced for reference in Fig. Appx. 1.2.

Bibliography

BAILEY, H. P. 1960, A method for determining the temperateness of climate; *Geografiska Annaler*, Vol. 42, p. 1–16.

BUDYKO, M. I. 1956, *The Heat Balance of the Earth's Surface;* (Trans. by N. I. Stepanova), U.S. Weather Bureau, Washington, 1958.

BUETTNER, K. J. 1962, Human aspects of bioclimatological classification; In *Biometeorology*, Vol. 1 (*Ed.* Tromp, S. W. and Weihe, W. H.) (Pergamon Press, London), p. 128–140.

CARTER, D. B. 1954, *Climates of Africa and India according to Thornthwaite's 1948 classification;* Publications in Climatology, Vol. 7, No. 4, Laboratory of Climatology, Centerton, New Jersey.

CHANG, J.-H. 1959, An evaluation of the 1948 Thornthwaite classification; *Ann. Assn. Amer. Geog.*, Vol. 49, p. 24–30.

CROWE, P. R. 1957, Some further thoughts on evapotranspiration: a new estimate; *Geographical Studies*, Vol. 4, p. 56–75.

FLOHN, H. 1950, Neue Anschauungen über die allgemeine Zirkulation der Atmosphäre und ihre klimatische Bedeutung; *Erdkunde*, Vol. 4, p. 141–162.

FLOHN, H. 1957, Zur Frage der Einteilung der Klimazonen; *Erdkunde*, Vol. 11, p. 161–175.

GENTILLI, J. 1958, *A Geography of Climate;* (Univ. of Western Australia Press), p. 120–166.

GREGORY, S. 1954, Climatic classification and climatic change; *Erdkunde*, Vol. 8, p. 246–252.

HARE, F. K. 1951, Climatic classification; In *London Essays in Geography* (*Ed.* Stamp, L. D. and Wooldridge, S. W.), (Longmans, Green & Co., London), p. 111–134.

MILLER, A. A. 1951, Three new climatic maps; *Trans. Inst. Brit. Geog.*, No. 17, p. 13–20.

SHEAR, J. A. 1966, A set-theoretic view of the Köppen dry climates; *Ann. Assn. Am. Geog.*, Vol. 56, p. 508–515.

SIBBONS, J. L. H. 1962, A contribution to the study of potential evapotranspiration; *Geografiska Annaler*, Vol. 44, p. 279–292.

THORNTHWAITE, C. W. 1933, The climates of the earth; *Geog. Rev.*, Vol. 23, p. 433–440.

THORNTHWAITE, C. W. 1943, Problems in the classification of climates; *Geog. Rev.*, Vol. 33, p. 233–255.

THORNTHWAITE, C. W. 1948, An approach towards a rational classification of climate; *Geog. Rev.*, Vol. 38, p. 55–94.

THORNTHWAITE, C. W. and HARE, F. K. 1955, Climatic classification in forestry; *Unasylva*, Vol. 9, p. 50–59.

THORNTHWAITE, C. W. and MATHER, J. R. 1955, *The Water Balance;* Publications in Climatology, Vol. 8, No. 1, Laboratory of Climatology, Centerton, New Jersey, 104 pp.

THORNTHWAITE, C. W. and MATHER, J. R. 1957, *Instructions and Tables for Computing Potential Evapotranspiration and the Water Balance;* Publications in Climatology, Vol. 10, No. 3, Laboratory of Climatology, Centerton, New Jersey, 127 pp.

TROLL, C. 1958, Climatic seasons and climatic classification; *Oriental Geographer*, Vol. 2, p. 141–165.

APPENDIX 2

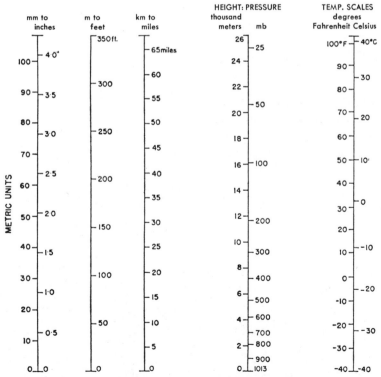

The left-hand scale of metric units can be read against those for inches, feet, or miles

Fig. Appx. 2.1.   Nomograms of height, length, and temperature.

APPENDIX 3

The International Metric System (SI Units)

Joule (J) is the SI unit of energy.

   1 gram-calorie = 4.1868 J.

Watt (W) is the SI unit of power.

   1 watt = 1 joule/sec.

   1 cal/cm²/min = 0.6975 kW/m², 0.06975 W/cm² or 69.75 mW/cm²

   (1 kW = 10^3W; 1 mW = 10^{-3} W)

Newton (N) is the SI unit of force. A force of 1 N accelerates a mass of 1 Kg by 1 m/sec².

   1 millibar = 100 N/m².

Subject Index

Absorption: of solar radiation, 2, 5, 15–28, 257; of terrestrial radiation, 2, 5, 15, 29–30, 257, 280

Adiabatic: chart, 65; lapse rate, 64–66, 91–92; motion, 221; process, 55, 64, 125, 249

Advection, 33, 35, 41, 143: fog, 128, 202–203; of heat, 33; of moisture, 54, 82, 136

Aerogram, 66–67

Aerosol, 1, 38, 56

Air: composition of dry, 1–6; density, 1, 3, 7, 40; volume, 1, 2, 7, 55

Airflow patterns, 180–188, 279–280: climatic characteristics of, in British Isles, 182–185

Air masses, 119–129, 147, 179, 180, 218, 219, 229: characteristics over Britain, 121, 124, 181–185; characteristics over North America, 121–124; climatic classification using air masses, 180, 302; general characteristics, 67, 120–124; mixing of, 55–56; modification of, 124–129

Air motion, 83–118: horizontal, 83–94, 96–100, 106–110, 111–114, 265; vertical, 41, 104–105, 111–114, 227

Albedo, 21, 197, 252, 262

Aleutian low, 101, 197, 279

Anabatic wind, 105

Anafront, 131–133

Andhis, 235

Angular: momentum, 110, 116, 227–228; velocity, 84, 105

Anticyclones (highs), 41, 100–104, 114–115, 182, 187, 188–191, 214, 227, 236–237: blocking, 188–191, 214; cold, 101, 104; Mackenzie winter, 101, 195; Siberian winter, 101, 178, 201, 208; subtropical, 20, 33, 97, 100–101, 103–104, 106, 112, 114–115, 117, 122, 123, 128, 139, 141, 213, 215–216, 219, 222, 226, 228, 231, 232, 236, 242, 247, 280; warm, 104, 236–237

Arctic air masses, 120–122, 124, 126, 127, 135, 140, 146, 182, 184, 185, 196

Arctic front, 138, 140, 179, 196

Aridity, 297–299, 301: index, 210, 298, 301; arid and semi-arid regions, 35, 215–217, 239, 240, 252

Aspect, effect on isolation, 27

Atmosphere: composition, 1–6, 40–41, 256–259; mass, 7–10, 38; mean temperature, 10; pressure (See Pressure); upper atmosphere, 40–41; vertical structure, 37–42

Atmospheric electricity cycle, 72

Atmospheric gases, 1–6: carbon dioxide, 1, 4–6, 29, 280–281; helium, 2, 41;

hydrogen, 2, 41; nitrogen, 1, 2, 9, 40, 41; oxygen, 1–3, 9, 40, 41; ozone, 2–5, 15, 28, 37, 39, 280; variations with height, 2–3; variations with latitude and season, 4; water vapor (See Water vapor)

Atmospheric pollution, 254, 256–259

Aurora, 40

Averages, 247–248

Azores high pressure, 100, 139, 178, 279

Bai-u season, 242

Baroclinic atmosphere, 124, 148

Baroclinic zone, 131, 138, 143, 144, 147, 194

Barometer, 7–8

Barotropic atmosphere, 119, 131, 148

Bergeron Theory of Precipitation, 61–63

Black body radiation, 10–11, 29

Blizzard, 206, 235

Blocking, 188–191, 214, 216

Bora, 214

Boreal forest, 196

Bowen's Ratio, 48

Boyle's Law, 7

Breeze, land and sea, 93–94, 217, 219, 231, 244–246, 247–248

Buran, 235

Calorie, 10, 308

Carbon dioxide, 1, 4–6, 29, 280–281

Centrifugal force, caused by Earth's rotation, 87

Centripetal acceleration, 83, 87–88, 227

Charles's Law, 7

Chinook, 91–93, 206

Circulation: cell, 111–113; general, 110–118; meridional, 111–112; zonal, 113

Climate: classification of, 47, 297–304; continentality, 12, 24, 27, 30, 179–180, 201, 204, 302; forest, 261–270; Mediterranean, 213–215, 302; mountain, 27, 42, 79–80, 91–93, 192–193, 199–200, 246–247; oceanicity, 179–180, 201–204; urban, 253–261

Climatic change: causes of, 279–281; post-glacial, 275–277; recent, 275, 277–279

Climatic types, 25, 43, 247, 298–304

Cloud, 58–60, 131–135, 147: associated with frontal cyclones, 59, 131–135; associated with hurricanes, 225, 227; base, 59, 80; classification, 58–60; droplets, 29, 57, 69, 70, 244; effects on radiation, 15, 17–18, 28–29; noctilucent, 40; towers, 59, 222, 224, 227, 229; warm clouds, 62, 80; cloud seeding, 62, 252

Geographical Index

Author Index